GOING PRIVATE

GOING PRIVATE

The International Experience with Transport Privatization

José A. Gómez-Ibáñez
and
John R. Meyer

The Brookings Institution
Washington, D.C.

About Brookings

The Brookings Institution is a private nonprofit organization devoted to research, education, and publication on important issues of domestic and foreign policy. Its principal purpose is to bring knowledge to bear on current and emerging policy problems. The Institution was founded on December 8, 1927, to merge the activities of the Institute for Government Research, founded in 1916, the Institute of Economics, founded in 1922, and the Robert Brookings Graduate School of Economics, founded in 1924.

The Institution maintains a position of neutrality on issues of public policy. Interpretations or conclusions in Brookings publications should be understood to be solely those of the authors.

1775 Massachusetts Avenue, N.W., Washington, D.C. 20036

Library of Congress Cataloging-in-Publication data

Gómez-Ibáñez, José., 1948-
Going private : the international experience with transport privatization / José Gómez-Ibáñez, John R. Meyer.
p. cm.
Includes bibliographical references and index.
ISBN 0-8157-3178-7 (alk. paper) : — ISBN 0-8157-3179-5 (pbk. : alk. paper)
1. Urban transportation—Case studies. 2. Bus lines—Case studies. 3. Express highways—Case studies. 4. Privatization—Case studies. 5. Contracting out—Case studies. I. Meyer, John Robert. II. Title.
HE308.G66 1993
388.4—dc20 93-29153
CIP

9 8 7 6 5 4 3 2 1

The paper used in this publication meets the minimum requirements of the American National Standard for Information Sciences—Permanence of paper for Printed Library Materials, ANSI Z39.48-1984.

Authors' Acknowledgments

THIS BOOK IS THE PRODUCT of a genuine and equal collaboration. Each author had an important hand in every chapter, and the order of authorship was determined alphabetically. If the credit is shared, so too are our many debts.

Our debts are numerous, moreover, since in this book we summarize and synthesize work we have done on a dozen projects over a six-year period. We have tried to acknowledge the most important institutions and individuals, but listing them all would be impossible and we apologize to those overlooked.

For financial support we are particularly indebted to the U.S. Department of Transportation and Harvard University. Various agencies and programs of the U.S. Department of Transportation sponsored the research for many of the case studies summarized in this volume. Our studies of the British experience with urban bus privatization and of U.S. private rail transit were funded by the Federal Transit Administration (FTA). The Federal Highway Administration (FHWA) supported our studies of Mexico's private toll road program. The Department of Transportation's University Transportation Center (UTC) research program supported our studies of U.S., French, and Spanish private toll roads. Harvard University provided sabbatical support to pull these many studies, and others, into a single volume. Some of our insights were also gained in work on privatization proposals for the governments of Indonesia, Bolivia, Chile and Spain, and for the World Bank.

Our debts to colleagues who participated in the research or criticized our drafts are also numerous. Arnold M. Howitt and Alan D. Wallis played such a major role in the studies of U.S. private rail transit proposals summarized in chapter 13 that we have listed them as coauthors of that chapter. Arnold Howitt also contributed greatly to an analysis of the political economy of subsidies, as summarized in the last third of chapter 14. Paul Kerin and Leslie Meyer Dean were our able collaborators in the study of British bus privatization summarized in chapter 4. In our studies of U.S. private toll roads (chapter 10), Marcella Butler provided important research assistance with the California proposals, and Laura J. Carter with the Virginia proposal. Johan Bastin assisted us in analyzing privatization projects in Indonesia, and Fernando C. Navarro with projects in Chile. Finally, Clell Harral of the World Bank and European Bank for Reconstruction and Development has helped educate us about privatization in nonmarket economies and former communist countries.

All these research projects would have been impossible without the assistance of the many public officials and private entrepreneurs who were generous with their time and patient with our many requests for information and assistance. Although it is impractical to list all who have been helpful, a few should be mentioned. In the United States we are particularly indebted to Stephen A. Lockwood, former associate administrator of the FHWA; Donald Symmes and Siegbert Schacknies of the International Programs Office of the FHWA; Ralph Stanley, former administrator of FTA and former CEO of the Toll Road Corporation of Virginia; William Reinhardt, editor of *Public Works Financing*; Maureen Gallagher, research director of the International Bridge Tunnel and Turnpike Association; and Carl B. Williams, assistant director of Caltrans. In France we owe special debts to Eric Boiteux of the Highways Directorate, Ministry of Equipment; Henri Cyna, the president of Cofiroute; Daniel Tannenbaum, president of SANEF; and Robert Lafont, general director of the Association of French Autoroute Societies. Our special thanks in Spain to Alexandro Sanchez Melero of the Ministry of Public Works and Transport; Carlos Martin Placencia, delegate of the government for Autopistas Concessionaires; and Jose Luis Ceron Ayuso, chairman of Autopistas del Mare Nostrum. In Britain we would like to acknowledge Peter Stonham, former editor of *Bus Business*; William Tyson, an economic consultant to the bus industry; Harry Blundred, the general manager of Devon General; Peter Sephton, general manager of South Yorkshire's Transport; and James Issac, general manager of West Midlands Travel. In Mexico, our special thanks to Jaime Luna Trail, director of planning for infrastructure at the Ministry of Communications and Transport.

We are also indebted to Sue Wood and Diana Murray, who did a wonderful job of preparing the manuscript and never complained about our frequent revisions. Sue Wood and Janet Gulotta also prepared many of the original research reports on the case studies.

Parts of three of the chapters have appeared in journal articles by the authors: an early version of chapter 4 was published in the *Journal of the*

American Planning Association, an early version of chapter 3 appeared in the *International Journal of Transport Economics*, and parts of chapter 14 appeared in *Traffic Quarterly*. Some of the themes of the book were also first developed in an article that appeared in the *Journal of Transport Economics and Policy* and in a book published by the Federal Reserve Bank of Boston.

We also benefited from many seminar presentations at Harvard and elsewhere. Among the many who furthered our understanding on such occasions are Robert Brown, John Kain, Thomas Moore, Kenneth Small, and Alan Walters.

Finally, at the Brookings Institution, Theresa Walker edited the manuscript, Laura Kelly verified it, and Rhonda Holland prepared the index. The views expressed in this book are ours alone and should not be ascribed to any of the persons or organizations mentioned above.

Contents

Tables

Figures

Chapter 1. Introduction

IN THE 1980S MANY COUNTRIES turned to private sources to provide services formerly offered by public agencies. Europeans, particularly the British and the French, were leaders in this movement. Europe had the greatest number of state enterprises to sell back to private investors, since it had nationalized many industries that had remained in the private sector elsewhere in the industrialized world. Developing countries also experimented extensively with privatization in the 1980s, with varying enthusiasm and effectiveness. Finally, political developments in the former Soviet Union and Eastern Europe at the end of the 1980s made privatization a worldwide phenomenon, popular with a remarkable range of governments. Transport was no exception to this pattern; indeed, in many ways it was at the cutting edge.

Varieties of Privatization

Privatization can assume many different forms, but three are most common: the sale of an existing state-owned enterprise; use of private financing and management rather than public for new infrastructure development; and outsourcing (contracting out to private vendors) public services previously provided by government employees. A wide variety of competitive, regulatory, and subsidy policies has accompanied these forms of privatization.

The first type of privatization, the sale of state-owned enterprises, has occurred often in Europe, in the developing world, and in former planned economies like those of Eastern Europe and the former Soviet Union.[1] In Western Europe, state enterprises have often been of a large scale, such as public utilities, transportation, and heavy industry. In Eastern Europe and the former Soviet Union, small- as well as large-scale activities have been state owned. Among the developing countries, too, both small and

1. The United States and Canada, however, have not been exempt from such activities, as Conrail, Air Canada, and several other cases illustrate.

big state enterprises are found, with the breadth and extent of such public enterprises often reflecting the legacy of colonialism rather than any imbedded political orientations; often, no alternatives to government existed when the colonial authorities and their private partners left the country.

The second principal form of privatization, private rather than traditional public sector infrastructure development, has become quite popular, at least in an experimental sense, almost everywhere. In the United States, several private proposals were made during the 1980s to build highways, urban rail transit, sewage and water treatment plants, solid waste incinerators, and landfills. The private provision of infrastructure was popular elsewhere, though, well before its revival in the United States. Private toll roads, for example, accounted for a large percentage of total high-performance highways built in France and Spain since the late 1960s. In recent years, moreover, other European countries, after long relying on tax financing of high-performance roads, have begun to explore private toll roads (for example, Britain and Scandinavia). A most dramatic illustration of the use of private means to finance and develop a major infrastructure facility is provided, of course, by the channel tunnel linking northern France with southeastern England. Indeed, very high-cost facilities, such as tunnels and bridges, have long been financed by tolls, even if not always privatized, in much of the world. In the same fashion, high-performance highways in developing countries have commonly been financed by tolls, and increasingly by private companies rather than public authorities.

The third common form of privatization, the takeover of "conventional" public sector functions through contracting to private vendors, has been mainly, although not exclusively, a U.S. and British phenomenon. Numerous undertakings have been proposed for such privatizations, including waste disposal, transit operations, school lunches, sewage and water treatment, airports and airways, and prisons.[2]

Privatizations have been accompanied by diverse regulatory, deregulation, and competitive policy packages. For example, privatizations are sometimes linked with added or new regulation to prevent abuse of ostensible monopoly powers. Thus the British added new regulatory procedures when privatizing telecommunications and public utilities. Similarly, as chapter 3 explains, several developing countries initiated

2. See, for example, John D. Donahue, *The Privatization Decision: Public Ends and Private Means* (Basic Books, 1989).

new regulatory procedures when they privatized previously state-owned bus enterprises. The prospective development in the United States of private toll roads and other privatization initiatives has prompted a reconsideration and renovation of many regulatory procedures.

Privatizations have also been accompanied by deregulation. The British bus experiment, described in chapter 4, is a prominent example. A few developing countries, as reported in chapter 3, have also largely or even totally deregulated when privatizing some of their bus systems (for example, Sri Lanka and Chile). Privatization combined with deregulation has been proposed in several instances for rail enterprises in developing countries.

At the other extreme, privatization sometimes occurs with deliberate enhancement of monopoly. For example, British Air was allowed to acquire most of its major British competitors before going public with a stock issue in its privatization.[3] The privatization of British airports (as discussed in chapter 12) illustrates even more egregiously monopoly enhancement combined with privatization: in that case all the major international airports in the United Kingdom were essentially put in the hands of one private owner. Thus a policy of separate (and therefore somewhat competitive) ownership of different airports by different companies was deliberately eschewed. A policy of intentional monopolization has accompanied several privatizations of industrial enterprises in developing countries as well.

Motives and Objections

The three basic types of privatization usually arise from three somewhat different motives. For the sale of state-owned enterprises or contracting out, a primary motivation has been a widespread belief that the private sector is inherently more efficient than the public sector. A privately managed enterprise or a private contractor, motivated by the possibility of profit, may have stronger incentives to be more cost conscious, efficient, and customer oriented than a public enterprise. These efficiency gains, if real, should eventually reduce the cost to the taxpayer of supporting the former state-owned enterprises or the contracted services.

In contrast, infrastructure privatization is often motivated by a desire to tap new sources of funds to supplement the constrained resources of

3. One might argue that this acquisition did not reduce competition much, however, since the principal competition to British Air traditionally has been foreign carriers.

the public sector. Efficiency may still be claimed as an important advantage as the private sector is often thought to build infrastructure cheaper or faster than public counterparts. But usually the primary concern is that the public sector simply does not have the financial resources to build the infrastructure needed. Unlike many other government services, moreover, infrastructure can often be supported by charges levied on users. Privatization offers the potential for financing infrastructure without overt increases in taxes; in many ways privatization can take an activity off the political agenda. Privatization thus has particular appeal when the public sector faces considerable taxpayer resistance and is unable to expeditiously finance badly needed facilities or activities that the private sector might undertake for a profit.

The prospect of immediate financial gain to government applies primarily to the outsourcing of existing state-owned enterprises or infrastructure facilities. Governments, of course, can realize significant proceeds only if the to-be-privatized enterprise can generate operating income in excess of that needed to cover operating expenses and finance expected new investment needs. Even then, the expected future cash flow may not be large enough for the government to recover its original investment. But any recovery is often advantageous, especially when government tax resources are limited. And certain types of government enterprises or infrastructure, such as airports, are often sufficiently profitable to make their sale attractive to cash-strapped governments.

These efficiency and financing arguments for privatization, however, are not always convincing. Critics point out that a newly privatized state enterprise may have few incentives to be efficient or market oriented if it operates in a monopoly or uncompetitive market, for example, while contracting out may not reduce costs if the bidding process is rigged or not competitive. Critics also argue that private contractors may hold back on the quantity or quality of services they render unless their performance is monitored, and the costs of such monitoring could offset any savings in efficiency.

Nor does private provision necessarily increase investment in infrastructure. A public agency often is able to tap private capital markets by issuing bonds secured by revenues from infrastructure facilities, just as a private firm would; such a strategy would neither increase taxes nor raise general obligation debt. Financing infrastructure on private capital markets may also displace some other form of private investment since privatization does nothing directly to increase the pool of savings on which the private capital markets draw. In short, the argument that privatization increases investment in infrastructure is strongest when the public sector's access to private capital markets is restricted for some reason. Even then, infrastructure may often displace some other form of private investment.

Privatization creates winners and losers and thus considerations of equity. Indeed, the debates over cost savings and efficiency often fail to distinguish between savings that are net efficiency gains and those that represent transfers from one segment of society to another. In privatizing a state-owned enterprise or contracting out, for example, costs may be reduced in part by shedding excess staff or by cutting wages. If the excess staff members have no other prospects for employment, their release is not necessarily a savings for society as a whole because the workers are still not engaged in productive activity; rather the layoffs may be viewed largely as a transfer from the employees, who are now not paid, to taxpayers or customers of the enterprise, who now pay less in subsidies or prices, respectively. Conversely, if the excess staff members find jobs elsewhere, then society should be better off because the laid-off employees will finally be productively employed; in this case, the employees may be no worse off, presuming their pay rates and conditions of employment are not much different, while the taxpayers or customers of the formerly state-owned enterprise are better off (in not having to pay for unproductive labor).

Similar confusions between net efficiency gains and transfers sometimes arise in the private provision of new infrastructure. Much of the public policy debate over privatization of infrastructure in the United States, for example, has focused on the potential cost advantages of private ownership or operation, and on related attempts to modify the federal and state tax codes to allow private firms (who must pay taxes) to compete on an equal footing with public ones (who are tax exempt). To the extent that the principal difference between public and private provision is simply tax treatment, privatization primarily involves transfers rather than net efficiency gains. If private enterprises are generally more subject to taxes, then the users of a newly privatized facility may be worse off because they will have to pay higher prices to compensate for the tax payments, while federal and state taxpayers in general will be better off because their tax rates can be reduced or public services increased.

Privatization also raises other issues, such as the relative abilities of private and public enterprises to deal responsibly with environmental, safety, or other social consequences of the services they operate. Neighborhood opposition and environmental controversies are often as much of a barrier to the construction of new infrastructure facilities as lack of government financing, especially in developed countries. Some may fear, rightly or wrongly, that the profit motive will lead a private company to skimp on environmental and other such responsibilities, thus entangling the privatization debate with other sensitive issues of public policy.

The Focus on Transportation

Transportation is a sector in which governments around the world are heavily involved and thus is a natural focus of experiments in privatization. Transportation infrastructure, such as highways, ports, and airports, absorbs a large share of public sector capital investments in most countries. State-owned transportation enterprises, for example, bus lines, railroads, and airlines, are fairly common, especially outside the United States; often they are highly unprofitable, usually because they are alleged to be inefficiently operated and thus require heavy government subsidies. Of course, inefficiency is not the only possible explanation; government-owned enterprises may be required to render some highly unprofitable "social services" or be severely restricted in the fares that they can charge.

Within transportation, highways and buses are arguably highly important modes, especially for domestic (as opposed to international) traffic. Highways carry the vast majority of domestic passenger and freight traffic in almost every country outside the old communist bloc, and highway use is rapidly growing around the world. In most urban areas, buses are the most heavily used form of public transportation, and they often carry more riders than rail transit even in the cities that have extensive subway or streetcar systems.

Almost every country has experimented with highway and bus privatization, and thus a rich variety of international experiences can be analyzed and compared. International comparisons are instructive because the privatization measures adopted, as well as the market and political environments, may differ significantly. Highway traffic is generally lighter while bus ridership is heavier in developing countries, for example, which tends (all else equal) to make highway privatization harder but bus privatization easier in the developing world. Regulatory and financial institutions are often less sophisticated in developing than developed countries, which can alter the advantages and disadvantages of privatization programs.

An equally important reason for focusing on highways and buses, however, is that they encompass all three of the most common and basic forms of privatization. Bus transit privatizations often mean the sale of an existing state-owned enterprise or, almost as commonly, the contracting out of public services previously provided by government employees. Privatizing the provision of high-performance highways is almost by definition an exercise in the use of private (rather than public) financing and management for new infrastructure development. As a corollary, the primary motive for privatizing buses is usually the prospect of cost or efficiency savings, translatable into subsidy reductions, while the primary motive for privatizing highways is usually to find alternative sources of financing for needed new investments.

Basic differences between highways and buses affect the prospects for privatization in other fundamental and interesting ways. Buses are relatively labor intensive while highways are capital intensive, and bus capital is mobile while highway investments are not. These differences usually make maintaining effective competition a more difficult problem for private highways than private buses; the threat that a new competitor might enter the market is much more credible with buses since the minimum investment required is modest and the capital can be used in other markets if the challenge is unsuccessful. Siting and environmental issues are generally much more important for highways than for buses for obvious reasons, and thus the question of how well the private sector copes with environmental controversies and responsibilities is more relevant to highway than bus privatization.

Overview

This book's narrative of various experiences with the privatization of transportation is roughly arranged in ascending order of difficulty, by mode and country. Thus we begin with bus transit in part 1, since its privatization problems seem somewhat more manageable than those for highways. The experiences of the less developed countries, where bus privatizations have been relatively easy to implement, are addressed first; then we turn to Europe, where Britain has been a leader in privatizing buses; and we conclude with the U.S. experience, where bus privatization has thus far proved difficult even though it is often advocated. Similarly, in part 2, when discussing highway privatization, we start with European roads where privatization has proceeded, though unevenly, in several countries on an extensive scale (especially France and Spain) and is being proposed or investigated in several others. Following that, we take a look at developing countries, where many private roads have been opened since the late 1980s, and then we look at the U.S. highway experience, where several private highway proposals are in advanced stages of planning and development but only one has begun construction (as of late 1993). Finally, in part 3, we explore the far rarer attempts to privatize two other forms of transportation infrastructure: rail transit and airports. The basic obstacle to rail transit has been that, whether publicly or privately owned, rail is inherently difficult to finance from farebox revenues alone, especially in developed countries. Airports, in contrast, are often potentially highly profitable, but the possibility that privatization might lead to large increases in airport charges to some users often arouses serious opposition to private provision.

The transportation privatization experiences examined are sufficiently rich and diverse to permit some tentative generalizations about the prospects for privatization in transportation and elsewhere. In developing these generalizations and related hypotheses about the comparative advantages of the public and private sectors, we consider the conventional concerns of cost and financing, as well as the oft-neglected dimensions of siting, equity, income transfers, pricing, and government regulation. Indeed, as already suggested, we argue that these other dimensions are often at least as important as, and often more important than, cost and financing. Thus we assess the circumstances in which privatization is likely to succeed and, by implication, fail for political as well as economic reasons.

Five basic lessons or themes, supplemented by some important corollaries, emerge from our analysis. First, competition in the markets in which privatized firms buy and sell is highly desirable if privatization is to succeed. Competition is especially important in encouraging the cost savings or efficiencies that often motivate privatization. Furthermore, competition reduces the risk that private firms can arbitrarily raise prices or restrict supply, thereby generating a demand for government regulation to prevent a monopoly abuse. Competition is not indispensable to privatization—indeed, in certain situations, for example, highways, too much competition can create serious problems. Findings in favor of competition can be found in most of the case studies, however, and are prominent in those pertaining to privatization of bus transit presented in chapters 2 through 6.

Second, privatization is easier to effect, all else equal, when the efficiency gains from privatization are fairly large, that is, when the private sector is for some reason or another inherently more efficient than the public sector. Among other advantages, large gains in efficiency provide the means for buying out other problems, such as those arising from concerns about the environment or equity. Needless to say, large gains in efficiency also enhance the incentives to privatize. As a corollary, enhanced and innovative services are often as important as cost reduction in augmenting the attractiveness of privatization.

Third, privatization is easier to implement when there are not too many redistributions or transfers linked with the privatization. Transfers, or even uneven sharing of benefits among different groups, can be impediments. Advocates of privatization sometimes mislabel transfers as efficiency gains, but opponents of privatization seldom make this mistake. For example, loss of subsidized services is often connected with privatization, and this loss clearly will be perceived, especially by those who formerly used the subsidized services, as a partial offset to any gains derived from new services or products created by privatization and deregulation (as in Britain's bus experience discussed in chapter 4). In simplest terms, there obviously are fewer political frictions to impede privatization when there are not too many transfers or redistributions.

Fourth, privatization works best when associated with fewer controversial consequences such as environmental concerns or general opposition to economic development or growth. The complexities created by these issues are well documented for private highway development in the United States (chapter 10). They are also important, although to a lesser extent, in privatization of the bus systems, especially in less developed countries (chapter 3).

Finally, privatization is easier when the activity or service approximately covers its costs, neither requiring significant government subsidy nor generating significant surplus. The need for subsidy is not a bar to privatization. But the need for subsidies, even if sought from private sources rather than government, greatly complicates the effort to privatize by invariably extending the nature and scope of the political discussion. It is far simpler and easier if privatization can be financed strictly from available tolls or fares. The U.S. experiences with private toll roads (chapter 10) and private urban rail transit (chapter 13) illustrate this point. Although large profits or surpluses do not necessarily prevent privatization—indeed, may make it more attractive to cash-strapped governments—large profits are likely to be controversial with the users of the activity and stimulate fears of monopoly abuse (as in the case of airports described in chapter 12).

In short, privatization hinges on many political, social, and economic concerns. Privatization is usually not simply a matter of efficiency improvements or capital augmentation but also depends on such deeply imbedded societal concerns as equity, income transfers, environmental problems, and attitudes toward taxation and the role of government.

Part One
Privatizing Services: Urban Bus Transit

Chapter 2. The Privatization-Regulation Cycle

IN THE PAST TWO DECADES almost all developing countries have experimented with different forms of ownership and regulation of urban buses. In the developed world, Great Britain undertook ambitious and unprecedented experiments in bus privatization in the mid-1980s. Even the United States experimented in the 1980s as public bus authorities contracted with private companies to provide various services. Two basic considerations have driven this widespread experimentation: bus services are important, and they are almost universally subject to a degenerative regulatory or managerial cycle that periodically endangers their availability.

The Importance of Urban Bus Services

Urban transport is often divided into the human or foot-powered modes, primarily walking or the bicycle, and the motorized modes. The motorized modes can be divided into those vehicles owned and operated by private individuals, most notably the automobile, and those available to the public, though not necessarily publicly owned, such as the bus, taxi, streetcar, subway, or commuter railroad. Buses are the principal form of motorized public transportation in the cities of developed and developing countries, with bus ridership typically higher than that of all the other public modes combined.

Urban bus services are especially important in the cities of developing countries. The urban transport systems of developing countries are strained by rapid population growth and by the shift, as incomes grow, to higher-quality and more costly transportation modes. In the poorest cities the shift is from foot-powered modes to motorized public transport, while in the cities of the wealthier developing countries, citizens move from public transport to the private automobile. In both the poorer and the wealthier cities, however, the most important motorized modes are typically the conventional-sized bus (with thirty-five or more seats), the minibus (which has thirty seats or fewer, and often as few as ten to fifteen), and several close cousins to the minibus, such as the shared-ride taxi or jitney. Buses and minibuses often carry 50 percent to 80 percent of all motorized trips in the cities of developing countries (table 2-1). Good bus transportation is therefore important to meet the aspirations for a motorized mode in the poorer cities and to help forestall unnecessary growth in automobile traffic in the wealthier cities of the developing world.

Buses are important even in the cities of the developed world where the private automobile dominates urban transportation. The private automobile accounts for more than 90 percent of all motorized person trips in U.S. urban areas and between 60 percent and 80 percent of motorized urban trips in Britain and Western Europe.[1] But public modes are important in relieving congestion in the largest and densest metropolitan areas of the developed world, particularly during the peak or rush hours and to and from central business districts. And in many smaller cities people who do not have access to a private automobile, perhaps because they are too poor to own a car or too young or infirm to drive, must rely on public transit. The bus is the primary means of providing such services in developed countries. In the United States, for example, urban buses carry three times more passengers than the several urban rail modes (subway, streetcar, and commuter rail) combined, and almost twice as many as the taxi.[2] Similarly, in Britain urban buses carry four times as many passengers as urban rail and three times as many as the taxi.[3]

The Provision of Bus Services

The provision of urban bus services requires choices about ownership, government regulation or control over the business decisions of bus firms, and whether and how bus services should be subsidized.

The basic options for ownership are a public authority or a private firm. Hybrid forms are also possible. In West Africa several important

1. John Pucher, "Urban Travel Behavior as the Outcome of Public Policy: The Example of Modal-Split in Western Europe and North America," *Journal of the American Planning Association*, vol. 54 (Autumn 1988), p. 509.

2. John R. Meyer and José A. Gómez-Ibáñez, *Autos, Transit and Cities* (Harvard University Press, 1981), pp. 25, 28.

3. United Kingdom, Department of Transport, *Transport Statistics of Great Britain, 1976–1986* (London: Her Majesty's Stationery Office, 1987), pp. 106, 137, 146. U.K. Department of Transport, *National Travel Survey: 1978/9 Report* (London: Her Majesty's Stationery Office, 1983), p. 19.

TABLE 2-1. Mode Chosen for Urban Trips in Developing Countries, Selected Cities

Percent

	Motorized modes								
	Public modes			*Private modes*					
Region	*Bus*	*Minibus*	*Shared taxi or jitney*	*Normal taxi*	*Rail*	*Motorcycle*	*Car*	*All motor modes*	*Percent motorized/ percent by foot*
Africa									
Abidjan, Ivory Coast (1981)[a]	40	12	...	13	...	...	35	100	n.a.
Dakar, Senegal (1981)[a]	36	19	11		...	...	33	100	n.a.
Bamako, Mali (1984)[a]	2	22	7	...	...	41	29	100	37/63
Donala, Cameroon (1982)[a]	27	...	31	...	...	20	22	100	64/36
Yaounde, Cameroon (1982)[a]	16	...	33	...	...	1	49	100	45/55
Nairobi, Kenya (n.d.)[b]	30	...	30	...	...	...	34	100	n.a.
Cairo, Egypt (n.d.)[c]	70		15		...	...	15	100	n.a.
Tunis, Algeria (n.d.)[c]	61		1		10	...	24	100	n.a.
Far East									
Bangkok, Thailand (n.d.)[c]	55		20[d]		...	...	25	100	n.a.
Jakarta, Indonesia (n.d.)[c]	51		21[d]		1	27[e]		100	n.a.
Latin America and Caribbean									
Kingston, Jamaica (n.d.)[f]	66		n.a.[f]	n.a.[f]	...	n.a.[f]	20	100	71/29
Santiago, Chile (n.d.)[b]	42	23	10		10	...	15	100	n.a.

n.a. Not available.

a. Richard Barrett, *Urban Transport in West Africa*, Technical Paper 81 (Washington: World Bank, 1988), pp. 62, 108, 82, 41.

b. Alan Armstrong-Wright and Sebastien Thiriez, *Bus Services: Reducing Costs, Raising Standards*, World Bank Technical Paper 68 (Washington: World Bank, 1987), pp. 85, 89.

c. Alan Armstrong-Wright, *Urban Transport, A World Bank Policy Study* (Washington: World Bank, 1986), p. 47.

d. Approximately half are small three-wheeled or four-wheeled vehicles; see Armstrong-White and Thiriez, *Bus Services*, p. 88.

e. Includes small three-wheeled vehicles.

f. John G. Schoon, "Privatized Bus Service in Kingston, Jamaica: Perspectives on Service and Proposed Improvements," paper presented at the annual meetings of the Transportation Research Board, January 1990, Washington, pp. 2–3. Other modes, possibly including taxis and motorcycles, account for 6 percent of all motorized and foot trips.

urban bus companies are owned jointly by the government and private investors, for example, while in several U.S. cities local public authorities own the urban bus company's assets and employ the drivers but contract with private management companies to manage the operation.

Government regulation is an issue primarily where bus companies are privately rather than publicly owned or operated. The scope of regulation varies widely, especially in the developing countries. Regulations governing safety are common, such as requirements that vehicles be inspected for road worthiness or that drivers be licensed. In many cities local authorities also regulate entry and exit by requiring permission for a firm to initiate or abandon specific services or routes. The government may also set fares or establish standards for service frequency, hours of operation, and other aspects of the quality or quantity of service.

Subsidies to keep fares lower or services more extensive than they otherwise might be are an option whether the bus company is publicly or privately owned. The government may provide subsidies directly in the form, for example, of cash payments to offset the firm's deficits, exemptions from fuel or other taxes normally paid by other businesses, or the provision of government-purchased vehicles or facilities. Alternatively, the government may avoid direct expenditures by establishing a system of subsidies across different types of bus services, from profitable to unprofitable, commonly called cross subsidies.

The Cycle of Private and Public Involvement

Many cities in both developed and developing countries have experienced a fairly similar cycle of private and public involvement in urban bus service.[4] The cycle, which can be divided roughly into ten phases, begins with the private sector being largely responsible for urban bus service, followed by increasing public involvement and eventually, at least in some cases, by a partial or complete return to private provision (figure 2-1). Needless to say, the cycle is not always the same in all contexts or stages of development but does seem to occur in many different circumstances.

Phase one in the cycle is the entrepreneurial stage when the industry emerges. At this stage numerous individuals or small firms, often with

4. This section is drawn from an earlier publication of the authors: John R. Meyer and José A. Gómez-Ibáñez, "Transit Bus Privatization and Deregulation Around the World: Some Perspectives and Lessons," *International Journal of Transport Economics*, vol. 18 (October 1991), pp. 231–58.

FIGURE 2-1. Stages in the Transit Bus Privatization-Regulation Cycle

1. Entrepreneurial
2. Consolidation
3. Regulation of fares and franchises
4. Decline in profitability
5. Withdrawal of capital and services
6. Public takeover
7. Public subsidies
8. Declining efficiency
9. Dilemma of subsidy cuts, fare increases, and service cuts
10. Privatization

only a few vehicles in each of their fleets, provide the service. In some cities, particularly in the developed or industrialized countries, this phase occurred more than one hundred years ago, when horse drawn vehicles were the reigning technology, although often a resurgence of individual or small-firm entrepreneurship took place when the motor bus or minibus first appeared. In the newly developing countries, this entrepreneurial phase usually occurred much later, and in several instances even holds today.

The second phase is characterized by mergers and consolidation of small operators into a few dominant associations or companies, often with little overlap between their route networks. In cities that were large and sufficiently wealthy around the turn of the century to support horse drawn or electric street railways, the emergence of the railway technologies was often partly responsible for this consolidation. Those cities that bypassed the street railway phase, or where entrepreneurship reemerged once the motor bus developed, usually experienced consolidations and mergers as well, although often at a later date.

Consolidation is typically followed by phase three in which the government regulates fares and issues franchises to the firms (to control routes and entry). Fare regulation and franchising are often a response to the perceived or real market power of the newly consolidated and larger firms created in phase two. In other cases regulation may be a response to a period of general inflation in the economy and the public outcries against consequent fare increases on services that had become popular and on which the city felt dependent. Others argue for regulation to eliminate so-called chaotic or destructive competition. Too many competitors have appeared, the regulators say, with unfortunate consequences for service stability, safety, reliability, and so on. Indeed, when this destructive competition is alleged, regulation may precede or be simultaneous with consolidation, reversing or compressing the timing of phases two and three; sometimes regulation may even be a prerequisite to consolidation by creating the market stability needed to finance consolidation.

In phase four a gradual downward trend appears in the profitability of the private firms followed by phase five, capital withdrawal and service reductions. Profitability usually declines because government regulators are reluctant and slow to allow fares to increase during periods of cost escalation generated by numerous causes, including general inflation, increasing urban congestion, or productivity increases not keeping up with wage increases in this labor-intensive industry. In the developed countries, profitability often declines because of the rapidly increasing popularity of the automobile and the reluctance of regulators to allow the firms to prune routes and adjust service accordingly. Whatever the cause, service is soon being provided in aging and often dirty or otherwise undermaintained vehicles. In third world cities where demand for public transport is increasing, the supply of vehicles becomes inadequate so that existing routes are overcrowded and service is not extended to outlying growth areas.

In phase six, the public authorities usually take over the ailing private firms followed by phase seven when an infusion of public subsidies restores capital and services. Public subsidies are commonly viewed as the only option for maintaining or expanding services at reasonable fares, and public subsidies are often deemed more acceptable when accompanied by public ownership. Public assistance is usually intended to expand service to accommodate increasing demand in developing countries, while stabilizing services in the face of stagnant or falling demand is more often the motive for subsidies in the already industrialized countries.

Public takeovers and subsidies are usually followed by increasing operating inefficiencies and rising real unit costs in phase eight. Often wages for bus company employees increase steadily and more rapidly than other comparable wages (or inflation), labor productivity declines because of overstaffing, and vehicle utilization declines because of inadequate attention to maintenance or slack scheduling practices. Eventually costs rise so much that bus subsidies become a significant burden on the public treasury. In phase nine, public authorities face an increasingly difficult and unattractive choice between further increases in public subsidies or significant and unpopular fare increases or service reductions.

In some cities, this dilemma is resolved by entering phase ten. The responsibility for bus service is returned once again to the private sector. Usually this shift is accompanied by a sharp reduction or outright elimination of public subsidies. Sometimes these subsidy reductions are based on the hope that the greater efficiency of the private sector will be enough to offset the loss of subsidies without forcing fare increases. In most cases, public regulation of fares is retained, in effect restarting the cycle in phase three. In a few notable exceptions, specifically Britain and a few cities in the developing world, fare regulation is also eliminated and the cycle returns to its first or second phase.

Differences between Developing and Developed Cities

Several important differences occur in how the cycle of private and public responsibility evolves in developed and developing countries. To start, the dissatisfaction with inadequate service by regulated private companies is not the cnly, or even the primary, motivation for public takeovers in some developing countries. In some cases, public ownership is an important political symbol, of independence from former colonial regimes or of the socialist ideology of the government. Public takeovers in many developing countries coincided with independence from colonial rule, especially in India and Africa where urban bus services in the largest cities were often provided by monopoly franchises granted to European companies.

The desire to modernize transport technology and economize on scarce street space also contributed to the push for public takeovers in some developing countries. In both developed and developing cities, the consolidation of numerous private bus firms into a single public authority commonly was promoted as a source of greater efficiency by rationalizing route networks and eliminating redundant services. In some cities in developing countries, however, especially in Asia, public ownership was motivated by the desire to hasten the replacement of smaller public transport vehicles (for example, pedacycles, motorized tricycles, jitneys, or minibuses) with modern and larger conventional buses that would supposedly use congested street space more efficiently. Critics argued that pedacycles and motorized tricycles impede traffic because they are slow; both they, the jitneys, and even minibuses require more street space per available seat, and possibly per passenger carried, than the modern standard-sized bus.[5] Whatever the merits of these arguments, moreover, elimination of pedacycles and motorized tricycles was consistent with the modern image that many developing countries wished to project to the outside world.[6]

Commonly, too, the argument, to a certain extent spurious, has been made in developing countries that the choice of transport technology dictated the form of ownership. In some cases it was argued that small private operators would be unable to raise the capital needed to purchase modern buses, even though private operators of conventional buses often were already in existence and their expansion might have been encouraged by a more generous regulatory environment or better franchise terms. Private operators were sometimes alleged to operate dangerously by racing competitors to stops, although this practice too conceivably could have been remedied by exclusive franchises on individual routes or more stringent enforcement of traffic regulations.

A more fundamental difference between developed and developing countries' public transit experiences is that the cycle of private and public responsibility for bus service is more compressed in the developing than developed countries. In large part this timing difference occurs simply because buses carry a much larger portion of urban trips in developing countries.

The high usage of buses increases the political temptation to intervene, whether or not there is an objective need to do so. There is enormous political pressure to keep bus fares low simply because buses are the majority mode of urban transportation. However, it is also risky to keep bus fares too low. To start, artificially low fares may induce inefficient patterns of urban development, such as unduly low densities of settlement. Direct government subsidies to support low bus fares could easily absorb a large share of the local public budget because buses are the principal mode of urban transportation. And overly stringent fare controls, without direct government subsidies, could easily leave the bus firms, whether public or private, without sufficient resources to provide this essential public service. As a consequence, the public takeover phase

5. Alan Armstrong-Wright, *Urban Transport: A World Bank Policy Study* (Washington, D.C.: World Bank, 1986), p. 25.

6. Perhaps a greater concern, though seldom voiced, is the political danger of critical urban services in the hands of numerous single-vehicle or small entrepreneurs, who may be difficult for the government to control.

is sometimes done less completely, and more experimentation occurs with mixed systems in the developing than the developed countries.

In essence, the developed countries can afford the luxury of relying on publicly owned and heavily subsidized urban buses more easily than their less developed counterparts. Urban bus companies have been publicly owned in virtually every major city in North America and Europe since the late 1960s.[7] Public ownership was rapidly accompanied by heavy government subsidies. Approximately 60 percent of bus operating costs and virtually all capital costs are paid by government subsidies in the United States, for example, while many European countries subsidize over 50 percent of bus costs.[8] Although these subsidies are often justified as means of controlling auto congestion or of providing an option for households without an automobile, a large part of the aid has been absorbed by growing inefficiencies rather than expanded services or reduced fares.[9]

By contrast, the developing world, of necessity, has had to be more imaginative and innovative. Private ownership is much more common, as are mixed regimes where some companies are public and others private. Subsidies are much rarer and less substantial, even though the various rationales (of discouraging automobile congestion or aiding the poor) are probably at least as plausible in the developing as the developed world. The extensive innovation and experimentation in the cities of the developing world make their experiences with urban bus privatization fascinating and instructive.

7. There are a number of useful histories of public involvement in the urban bus industry and its predecessor, the street railway, including Edward S. Mason, *The Street Railway in Massachusetts: The Rise and Decline of an Industry* (Harvard University Press, 1932); Richard Dolomon and Arthur Saltzman, "A Historical Overview of the Decline of the Transit Industry," *Highway Research Record*, no. 417 (1972), pp. 1–11; and David W. Jones, Jr., *Urban Transit Policy: An Economic and Political History* (Prentice Hall, 1985), pp. 28–95.

8. Pucher, "Urban Travel Behavior," p. 511; and F. V. Webster and others, "Changing Patterns of Urban Travel Behavior," *Transport Reviews*, vol. 6 (January–March 1986), pp. 49–86.

9. FSee, for example, Don H. Pickrell, "Rising Deficits and the Uses of Transit Subsidies in the United States," *Journal of Transport Economics and Policy*, vol. 19 (September 1985), pp. 281–98.

Chapter 3. Developing Countries: A Diversity of Experiences

TOTAL RELIANCE ON publicly owned bus companies is relatively rare among the cities of the developing world, and even cities that do have extensive public bus service often face competition from motorized tricycles or other forms of privately owned transportation offering services similar to that of a bus (as in India).[1] At the other extreme, urban bus services can be provided entirely by the private sector, as occurs in many of the smaller cities of Latin America and Asia. Even then, however, the local public authorities will specify fares, the routes that firms can serve, and sometimes the number of vehicles or the frequency of service on the routes.

By far the most common scheme for providing urban bus services in developing countries is a mixed system of publicly and privately owned buses. Often the conventional full-sized buses are operated by a publicly owned corporation while private operators provide significant minibus services. Sometimes the public sector dominates, as in Dakar, Senegal, where the subsidized public bus company is the exclusive provider of bus services within the city, while private and unsubsidized minibus operators are restricted to the suburbs and their fares and licenses tightly controlled. Sometimes the private sector dominates; in Accra, Ghana, for example, the public bus companies carry only 20 percent of all public transport

1. India undertook public ownership after independence from British rule, and public bus companies now supply all bus services in all but one or two of India's largest cities. The principal exceptions are Calcutta, where about one-third of the conventional bus services are publicly operated and two thirds privately operated, and Delhi, where the public transport firm contracts with private bus companies for some services. In many Indian cities, however, privately operated pedacycles and motorized tricycles carry as many or more passengers as the subsidized public bus companies. P. R. Fourace and others, *Public Transport Supply in Indian Cities*, Report 1018 (Crowthorne, England: Transport and Road Research Laboratory, 1981), p. 6, table 5.

trips while the remaining 80 percent are carried by private firms operating conventional buses, minibuses, and converted trucks.[2]

Both the mixed and the fully privatized bus systems have often developed in response to the inadequate services provided by recently nationalized and deficit-ridden public bus companies, described by the cycle of privatization outlined in chapter 2. A common solution to these deficit problems is for the government to reverse its position and privatize transportation, or, more often, to allow another less regulated and privately operated mode, such as minibuses, to step in and fill the gaps that publicly owned service does not fill. In the latter situation, the emergent private companies may be allowed to charge a higher fare or may be so much more cost efficient than the public or conventional bus that they can make a profit at the old low fare.[3]

Three types of privatization reform are important. The first, and by far the most common, is privatization accompanied by some public regulation, particularly for fares. Two other types of reform are far rarer but extremely interesting: privatization without fare regulation and privatization accompanied by the maintenance of subsidies for unprofitable routes.

Privatization with Fare Regulation

Privatization with regulation has arisen from several circumstances, occurring most commonly when the shortcomings of public ownership become apparent even before the initial intended public takeover of bus operations is complete. In these cases, privatization has taken the form of abandoning, more or less in midstream, further public takeovers and retaining the old regime for regulating private operator fares and routes. This pattern is common in Asia and to a lesser extent in Africa. In Jakarta, for example, the public took over several large companies that provided service with standard-sized buses in the 1970s. The intention was to expand services, gradually replacing the privately owned but publicly regulated pedicabs, motorized tricycles, minibuses, and jitneys that filled Jakarta streets, or at least restrict them to feeder services on secondary streets or in outlying areas. Although pedicabs were successfully restricted, at least initially, the growth in the subsidies required by

2. Richard Barrett, *Urban Transport in West Africa*, Technical Paper 81 (Washington: World Bank, 1988), pp. 52–55, 108–11.

3. Alan Armstrong-Wright and Sebastien Thiriez, *Bus Services: Reducing Costs, Raising Standards*, Technical Paper 68 (Washington: World Bank, 1987), pp. 81–84.

the Jakarta public bus company forced the government to slow or stop its program of restricting private motorized tricycles and minibuses (which still carry the majority of the city's public transport trips), and to leave fourteen bus companies in private hands.[4]

Sometimes privatization takes place after a planned public takeover of all bus services is completed and the old regime for regulating private fares and routes dismantled. In some cases, the public company reverts to the private sector (for example, Buenos Aires, Argentina, or Kingston, Jamaica).[5] At other times the public bus company is retained but private bus companies are allowed to compete (for example, Casablanca and Rabat, Morocco).[6] Whether the services have been fully privatized or mixed, however, a system for regulating private fares and routes has usually been reimposed.

Costs and Fares

Privatization has often allowed cities in developing countries to eliminate or check growth in subsidies while still maintaining or expanding services, largely because the costs of private bus companies are often much lower than those of their public counterparts. While public companies need not have high costs, and some clearly do not, public bus companies all too commonly have costs twice or so as high per paying passenger served as their private competitors.[7] A combination of excessive staffing, higher wage rates, lower vehicle utilization, and greater farebox revenue leakage usually accounts for these higher unit costs of the public sector. Private companies also often pay their employees less, particularly if they use minibuses, which usually have less stringent requirements for licensing of drivers. Public bus companies are rarely able to deploy more than 60 percent to 70 percent of their bus fleets during the

4. See Peter J. Rimmer, *Rikisha to Rapid Transit: Urban Public Transport Systems and Policy in Southeast Asia* (New York: Pergamon Press, 1986), pp. 160–68, and Michael W. Roschlav, "Nationalization or Privatization: Policy and Prospects for Public Transport in Southeast Asia," *Transportation Research A*, vol. 23A, no. 6 (1989), p. 414.

5. Armstrong-Wright and Thiriez, *Bus Services*, pp. 95-96, and Alan Armstrong-Wright, *Urban Transport: A World Bank Policy Study* (Washington: World Bank, 1986), p. 21.

6. John Damis, "Privatization of the Transport System of Rabat, Morocco," adapted by the staff of The Twenty-first Century Trust for use at a workshop on privatization, England, July 1988.

7. For example, Bombay and Madras are both thought to have relatively efficient public bus systems; see Armstrong-Wright and Thiriez, *Bus Services*, pp. 21, 90–93.

peak hours while private operators often deploy 80 percent to 90 percent. Finally, fare evasion by passengers and theft by bus crews seems more prevalent among public firms (thereby reducing both the actual and the reported revenue from passengers served).[8]

Because of lower costs, privatization often does not lead to pressures on regulators to increase fares. Indeed, the most convincing demonstrations of the lower costs of private operators is that public operators often incur large losses while private operators make a profit in those cities where bus services are provided by both types of operators and both charge the same regulated fares. In Jakarta, for example, both the public and one private firm operate conventional buses while the remaining private firms operate minibuses. Both the public and all the private firms charge the same fares, but the public firm needs a 50 percent subsidy while all of the private firms make a profit; similar situations exist in Accra and Calcutta.[9] In a few cases, such as Khartoum, a private operator of conventional buses apparently charges even less than the public operator.[10]

Privatization has led to higher fares in some cases, however, though accompanied by expansions of service. In Rabat, Morocco, and Kingston, Jamaica, for example, fares were held so low that government regulators felt they had to raise them when they opened the market to private operators.[11] In the few cases where privatization has been accompanied by deregulation, as discussed more fully later, fare increases have also occurred.

Safety and Congestion

The extent to which privatization has led to increased traffic safety and congestion problems is unclear. Reports of aggressive driving by highly motivated private operators are fairly common, particularly because in many cities a large number of small firms or individual owner operators will ply the same route. Some observers argue that better public provisions for on-street bus stops, off-street terminals, and the enforcement of

8. Armstrong-Wright and Thiriez, *Bus Services*, pp. 6–11.

9. Roschlau, "Nationalization and Privatization," p. 414; Armstrong-Wright and Thiriez, *Bus Services*, pp. 81–82, 87–88, 90-91; and Barrett, *Urban Transport in West Africa*, p. 53.

10. Armstrong-Wright and Thiriez, *Bus Services*, pp. 93–94.

11. Damis, "Privatization of the Transport System," and John G. Schoon, "Privatized Bus Service in Kingston, Jamaica: Perspectives on Services and Proposed Improvements," paper presented at the annual meetings of the Transportation Research Board, January 1990.

traffic regulations would ease the problem. In some cities the problem has been relieved by the development of route associations, which are organizations of very small private operators (usually with one to three buses each) that band together to provide common facilities (such as bus stations and maintenance depots) and sometimes coordinate schedules on their routes. Route associations may also act to limit competition in undesirable ways, however, such as by restricting entry or fixing fares.[12]

Privatization is often blamed for increasing traffic congestion by promoting the use of smaller vehicles. Of course, there is no inherent reason why private operators could not be required to use larger vehicles especially with a continuing regime of regulation. Private operators tend to select smaller vehicles when given the opportunity, however, because they are cheaper to operate, broaden the variety of services offered, and are usually popular with customers.[13] Smaller vehicles often can serve outlying areas where larger vehicles would be unprofitable or provide a frequency of service and a guarantee of a seat that the larger vehicles find difficult to match. Indeed, minibuses and similar vehicles are often so popular with the public that government planners have been forced to back down from efforts to suppress them, as in Hong Kong.[14]

Overview

On the whole, privatization with regulation seems to have benefited almost all parties in the many cities of the developing world where that policy combination has been pursued. Taxpayers have benefited because the burden of public bus subsidies has been reduced. Riders have probably benefited most of the time, especially when services expanded greatly with the introduction of private operators with little or no increase in fares. Finally, labor may not have lost much, or it may have gained because service expansions usually mean more jobs, though sometimes at lower wages.

The chief risks of privatization with regulation are twofold. First traffic congestion and safety issues arise, especially where the local government's ability to provide bus stops or terminals or to enforce traffic regulations seems limited. Second, the retention or reestablishment of public regulations, particularly for fares and entry to the industry, might in the long run seriously limit the ability of private operators to maintain

12. Armstrong-Wright and Thiriez, *Bus Services*, pp. 13–34.

13. Ibid., pp. 22–31.

14. The debate over minibus restrictions in Hong Kong is described in Rimmer, *Rikisha to Rapid Transit*, pp. 77–98.

or expand services, thus possibly renewing the cycle of public takeovers and subsidies. This contingency makes the examination of the few cases where privatization has been accompanied by fare deregulation of particular interest.

Privatization with Fare Deregulation

In some cities in the developing world certain types of public transport have always been privately operated and largely unregulated (except for requirements for vehicle inspection and driver licensing).[15] It is difficult to infer from these cases what the effects of deregulation might be, however, since no former regulatory regime or comparable existing regulated service provides a benchmark for comparison.

In the developing world the only two widely cited examples of privatization with deregulation are Colombo, Sri Lanka, and Santiago, Chile. The two cases are slightly different. In Colombo, the government in 1979 permitted private operators to offer bus services largely free of fare and route regulation but maintained the previous publicly owned bus company. In Santiago, a publicly owned bus company, which shared the market with regulated private operators, was disbanded in 1980 while government fare and route regulations covering the private sector were also relaxed. In neither city were subsidies for unprofitable routes made available to private operators, although in Colombo the public operator continues to receive large subsidies.

In Colombo, overcrowding and inadequate service by the public company led the government to turn to private operators. As of 1985, the public operator, the Central Transport Board (CTB), operated around 3,000 buses in the Colombo area, most with seated capacities of fifty-five passengers each, while the then-new private operators deployed 4,000 buses, most with seated capacities of ten to thirty passengers each.[16] The

15. Certain minibus operators are largely free to get their own fares and routes in Hong Kong, for example, although conventional bus fares and routes and taxi fares are regulated. In Nairobi minibus fares and entry have always been largely unregulated, and in some cities of Nigeria fares for all types of bus services have not been controlled.

16. These figures are from Louis Berger International, Inc., East Orange, N.J., *Buses*, vol. 3 of *Sri Lanka Transport Sector Planning Study*, final report prepared for the Interministerial Committee for Coordination and Planning of Transport (Colombo, Sri Lanka, January 1988), tables E3-1, E3-2. Others report 5,800 public and 3,500 private buses for the "Colombo metropolitan region" (which may be a larger geographic area) but do not specify the year; see Armstrong-Wright, *Urban Transport*, p. 28, and Armstrong-Wright and Thiriez, *Bus Services*, p. 39.

private operators were free to set their own routes and fares, but the CTB's policy of maintaining extremely low fares limited the private operators' ability to raise their own fares, which are only about 5 percent higher than the CTB's.[17] Consequently, some private operators resorted to overloading and did not serve some of the CTB's most unprofitable routes. Ridership increased greatly in the years after deregulation, owing to a combination of normal population and income growth and the big increase in capacity and bus frequencies stimulated by deregulation.[18]

In Santiago, the private sector offered several types of public transport prior to the reforms, including services with thirty-five-seat "microbuses," twenty-seat "taxibuses," shared-ride taxis plying certain routes, and regular taxis. Microbus and taxibus fares were deregulated and restrictions on entry and routes relaxed in 1979, while taxi entry restrictions were relaxed in 1978 and fares for shared-ride taxi service deregulated in 1981. At the same time the government gradually disbanded its publicly owned bus company, shrinking its fleet from 710 microbuses in 1978 to 44 in 1980 and none in 1981.

Santiago's reforms brought a dramatic expansion in public transport capacity but large fare increases for some forms of transport (table 3-1).[19] Between 1978 and 1984 microbuses increased by more than 50 percent, taxibuses nearly doubled, and taxis offering either shared-ride or regular services almost tripled. The considerable entry into bus service created bus traffic jams for several hours a day in downtown Santiago. Under the "guidance" of strong route associations, fares approximately doubled in real terms (that is, net of inflation) on the microbuses and taxibuses, even though fares on the shared and regular taxis remained roughly constant in real terms during this period. The fare increases were larger than the estimated increases in the running costs of a microbus or taxibus, which rose by only 20 percent during the period, largely because of a 100 percent real increase in the price of fuel. Significantly, ridership increased

17. Louis Berger International, Inc., *Buses,* p. 10.

18. There is no careful study that isolates the effects of deregulation from economic growth, ethnic strife, or other factors that undoubtedly influenced ridership during that period; see Louis Berger International, Inc., *Buses*, chap. 3. For a more pessimistic view of the Sri Lanka reforms see J. Diandas, "Bus Managers and Users Vis-à-vis Ownership, Regulation, Competition, and Systems," paper presented at the seventeenth annual summer meetings of the Public Transport Research Council, London, 1989.

19. J. Enrique Fernández and Joaquín de Cea, "An Evaluation of the Effects of Deregulation Policies on the Santiago Chile Public Transport System," Department of Transport Engineering, Pontifica Universidad Católica de Chile, 1985, pp. 1–12.

TABLE 3-1. Changes in Public Transport Fleets, Fares, and Patronage in Santiago, Chile, 1978, 1981, 1984

Public transport	*1978*	*1981*	*1984*
Fleet			
Microbuses	3,877	4,197	4,602
Taxibuses	1,558	2,222	2,703
Taxis	15,000	30,000	40,000
Fares (constant 1984 Chilean pesos)			
Microbuses	12	14	26
Taxibuses	16	18	28
Taxis	151	177	154
Shared taxis	61	71	66
Metro	12	19	19
Passengers (millions each year)			
Private microbuses	736[a]	808	604
Private taxibuses	264	358	346
Subtotal private bus	1,000	1,166	950
Metro	63	130	110
Subtotal bus and metro	1,063	1,296	1,060
Taxis shared and regular	n.a.	n.a.	500
Total	n.a.	n.a.	1,560

SOURCE: J. Enrique Fernández and Joaquín de Cea, "An Evaluation of the Effects of Deregulation on the Santiago, Chile Public Transport System," Department of Transport Engineering, Ponitificia Universidad Católica de Chile, 1985, p. 1, tables 1, 3, 5.

a. Does not include passengers on publicly owned buses.

dramatically on Santiago's taxis, whose fares did not increase, but fell on its microbuses or taxibuses, whose fares increased.[20]

Microbus and taxibus fares apparently increased rapidly because of the anticompetitive controls of their route associations while taxi fares were stable because there were no taxi route associations. Many of the private buses were operated by small entrepreneurs with only two or three buses each. Traditionally these small operators had pooled themselves in associations for specific routes, which financed the bus terminals and dispatched the vehicles on a common route and schedule. After deregulation, these route associations began to set fares as well and to war on new entrants who refused to join the association or who set their own fares. These measures seem to have been relatively effective, perhaps because the small size of the individual operators made it risky to

20. Estimating the effects of privatization and deregulation on ridership is difficult because the Chilean economy suffered a severe recession starting in 1982, shortly after the reforms.

challenge the associations' rules.[21] Furthermore, charging slightly lower fares was probably not a terribly effective competitive policy since few riders would choose to wait at a stop for the cheaper bus. The high fares encouraged capacity expansion, however, and a dramatic decline in average bus loadings.

Santiago has tried to solve the problems of excessive bus congestion and high fares in several ways, some successful and some not. During the 1980s, the government first imposed restrictions on the total numbers of microbuses that could enter the downtown and then decreed that vehicles over a certain age had to be retired. This policy had little apparent effect on fare inflation, probably because government restrictions on entry only made it easier for the route associations to control competition.[22] In 1991, however, the government began to experiment with competitively tendering exclusive franchises for routes to different groups of operators, a policy expected to force the associations to offer lower fares and to reduce capacity to win the route awards.[23]

The experiences of Colombo and Santiago suggest that if competition can be maintained, fare deregulation probably will not lead to large increases in fares or monopoly profits. In Colombo the CTB's low fares were clearly an important constraint on the pricing behavior of private operators, but no published reports of collusive or anticompetitive behavior have emerged. Santiago's experience of stable fares in the competitive shared-ride taxis and large fare increases among the collusive taxibus and microbus route associations offers direct evidence of the importance of competition.

Riders in both cities probably benefited from the reforms despite the fare increases, although the evidence is much stronger for Colombo than for Santiago. In Colombo, ridership increased significantly, apparently because the service increases were substantial and the fare increases moderate. In Santiago microbus and taxibus ridership fell despite large service increases because the route associations kept the fares so high that considerable excess capacity appeared. If Santiago's travelers gained from

21. Fernández and de Cea, "An Evaluation," p. 6, and Armstrong-Wright and Thiriez, *Bus Services*, p. 90.

22. Ian Thompson, "Urban Bus Deregulation in Chile," United Nations' Economic Commission for Latin America and the Carribean, Transport and Communications Division, Santiago, Chile, July 1990.

23. The results of this policy, which is opposed by the route associations, were not immediately clear; see Daniel Fernández Koprich, "La Modernización del Transporte Público de Santiago: 1990–1992," paper presented at the Seminario Internacional Autoridad de Control en el Sector Transporte, Buenos Aires, Junes 1992, pp. 10–12.

deregulation, they did so through the more competitive shared-taxi market, where substantial expansions in service combined with modest fare increases apparently led to much higher ridership.

Privatization while Maintaining Subsidies

Privatization and deregulation in the developing world are seldom accompanied by the maintenance of any significant direct public subsidy. Some modest aid may be given in the form of a waiver of fuel taxes or the public provision of bus stations, for example, but usually very few or none of the subsidies that the former public bus companies enjoyed are made available to private successors because avoiding the heavy and growing burden of subsidies is usually an important motivation for bus privatization. If direct subsidies are maintained in developing countries, often they are only given to a public bus company that is not disbanded but remains in competition with unsubsidized private operators.

Subsidizing public but not private firms in a mixed system may make sense in the rare places where the public firms charge a much lower fare or supply clearly different services than their private counterparts, as in Rabat and Casablanca in Morocco or in Colombo, Sri Lanka. In the two Moroccan cities, the publicly owned and subsidized operator provides standard service in large buses at a low fare while the privately owned operators provide a higher-quality service using minibuses with guaranteed seating at a fare that is twice as high. The public operators also offer heavily discounted monthly passes to students, which has seriously contributed to their financial problems, while the private operators are not required to do so. The distinction between deluxe and standard services gave the Moroccan government a way of maintaining standard service at an affordable fare for the lowest-income travelers while allowing the private sector to enter the bus business and greatly expand capacity that had been constrained by public sector deficits. Chronic overloading of buses, long waits at bus stops, and inadequate route coverage were reduced because the active bus fleets approximately doubled in the two cities after private service began in Casablanca in 1985 and Rabat in 1986.[24]

24. The active fleet in Casablanca increased from 452 public buses in 1982 to 320 public buses and 520 private buses by the end of 1987 and in Rabat from 176 public buses in 1982 to 150 public and 200 private in 1987. See Damis, "Privatization of the Transport System of Rabat," and Slobodan Mitric, "Urban Transport Enterprises: A Tale of Two Cities," draft report by the World Bank, Washington, 1989, pp. iii, viii.

In Colombo, as noted earlier, the public operator provides service in large buses at a low fare while the private operators provide service in both small and large buses at slightly higher and unregulated fares. Some of the benefits of the Colombo system are similar to those of the Moroccan scheme, since the entrance of the private carriers in 1980 greatly expanded capacity while the maintenance of the publicly owned and subsidized firm provides a slightly more affordable alternative service on some of the most unprofitable routes that the private operators have shunned.[25] Public subsidies in Colombo may offer additional benefits as a check on private fares, since (unlike Morocco) private fares are unregulated.

Even when services or fares differ, however, such subsidies have serious drawbacks. The main problem is that the publicly subsidized services may remain inadequate and overcrowded because of a combination of the high public operator costs and the inability or unwillingness of the government to foot the subsidy bill for service expansions. This situation has reportedly been less of a problem in Casablanca than Rabat, because the Casablanca public operator has risen to the challenge of privatization by improving productivity and cutting costs.[26] But in both Moroccan cities, the publicly operated bus fleet is smaller than it was before the private operators entered, and in both Morocco and Colombo reports of overcrowding on the more affordable services continue.[27] Although service has greatly improved for people able to afford the higher-priced privately operated services, the same cannot be said for the lowest-income travelers or those on routes not served by private operators.[28] Conceivably, of course, if the unsubsidized private operators were allowed to offer standard as well as deluxe services in Rabat and Casablanca, they could offer the same or greater capacity at a similar or only slightly higher fare than the public companies offer.[29] And in Colombo, the damage that

25. Armstrong-Wright, *Urban Transport*, p. 28.

26. Mitric, "A Tale of Two Cities," pp. 52–54.

27. The problem is most serious in Rabat where the public company operates deluxe as well as standard services and its standard fleet dropped from 149 buses in 1983 to 81 buses in 1986. See Mitric, "A Tale of Two Cities," p. viii, and Armstrong-Wright and Thiriez, *Bus Services*, p. 39.

28. Total bus ridership (standard and deluxe) dropped close to the time of privatization in both Rabat and Casablanca, but this reduction was partly because of large increases in standard fares shortly before privatization and a simultaneous recession in the economy; see Mitric, "A Tale of Two Cities," pp. viii, ix.

29. This is likely because the public sector companies reportedly remain relatively inefficient; Mitric, "A Tale of Two Cities," pp. 55–58.

subsidies do by restraining privately operated services may be greater than the benefit they create by providing a more affordable public alternative, especially since the fare differential is small.

Direct subsidies are occasionally given to private operators but usually only as temporary measures to ease transitions from one regulatory regime to another. In Medellin, Colombia, for example, government regulators held down fares for so long that private operators could not earn enough to replace, let alone expand, their aging fleet of buses, and services became more and more dirty, unreliable, and inadequate. The government finally authorized a substantial fare increase but only for new buses. The government also gave the operators of old buses a modest daily subsidy to encourage them to operate for a while longer at the old fare. The subsidy smoothed the transition to the new higher fares, was modest, and was limited to the short remaining useful life of the old vehicles.[30]

A few developing countries contract with private operators to serve routes that the public bus company finds unprofitable, but generally subsidies are not given so the problems of contract management and enforcement are not severe. Local authorities in Istanbul and Bangkok contract with private operators to serve routes that are unprofitable for their public companies, for example, but the private operators are happy to do so because they can make a profit at the same fare that the public operator charges. In both cases, the private operators actually pay the authorities for the privilege of operating the routes.[31]

In the rare places where a subsidy is preserved after privatization, the usual form is a cross subsidy to avoid any direct burden on the public purse. Furthermore, problems of accountability and control often seem less obvious or more workable with a cross subsidy. Different types of cross subsidies have been employed in Buenos Aires, Santiago, Daejong (Korea), Casablanca, and Rabat. In Daejong, for example, the city has sixty bus routes: forty are profitable and twenty unprofitable. The private operators are organized into associations, and these associations rotate among the routes weekly so that each operator takes a turn in providing profitable and unprofitable services.[32] Similarly, in Casablanca and Rabat, packages of routes, each including some considered profitable and some not, were initially awarded to the new private operators through a

30. Interview with Jeffrey Gutman, World Bank, March 21, 1990, in Cambridge, Mass.

31. Armstrong-Wright and Thiriez, *Bus Services*, p. 15.

32. Armstrong-Wright and Thiriez, *Bus Services*, p. 33.

process of competitive bidding; the winner of a package was obligated to offer service on all of its routes.

The choice between cross and direct subsidies can raise concerns about equity. Whether it is fairer for some bus riders to be subsidized by other bus riders rather than by the general taxpayer depends on particular perspectives and the situation. Cross subsidies may seem more equitable if many taxpayers are poor rural residents who do not enjoy the benefits of subsidized urban bus services, for example, while general tax revenues may seem fairer if the most lucrative bus routes are patronized by the poor or if the burden of general taxation falls mainly on the well-to-do.

Cross subsidies are also not free of administrative or regulatory problems, although these burdens are probably no more demanding than the accountability and control problems raised by direct subsidies. One special problem with cross subsidies, for example, is the potential anticompetitive effects of the exclusive franchises needed. Operators must be protected from competition on their lucrative services to generate the profits needed to subsidize unprofitable services. If more protection is given than needed, however, the firms may earn excessive profits or indulge in sloppy management or other forms of technical inefficiency.

The risk of excess profit with a cross subsidy can be far less marked where a route association rather than a single firm is granted the exclusive franchise. As long as the route associations permit new members to join, competition among the operators should dissipate any excess profits above and beyond those required to cross subsidize the unprofitable services (as in Daejong). For single firms, by contrast, even if the franchises are initially awarded on the basis of competitive bids, subsequently they may be extended or renegotiated through a more informal and less competitive process. Even then, government fare regulations may prevent excess profits being earned. In Rabat and Casablanca, for example, franchises were not rebid for quite a while, but operators probably did not earn excess profits because the government did not allow fare increases and imposed a new tax on the operators after the initial awards.[33]

Another troubling problem with cross subsidies is insuring that the less lucrative as well as the more profitable services are both provided. Franchisees have incentives to skimp on the less lucrative services, say, by cutting frequencies or hours of operation. Such problems have been reported in Rabat and Casablanca.[34] Indeed, the monitoring requirements

33. Telephone interview with Slobodan Mitric, World Bank.
34. Telephone interview with Slobodan Mitric, World Bank.

for cross subsidies usually differ little from those of all but the simplest direct subsidy and may be a serious strain on government departments. Furthermore, government officials operating a plan for cross subsidies may underestimate the need to monitor, since no direct public subsidies are involved.

Basic Lessons

While several lessons emerge from an examination of privatization of urban bus operations in developing countries, the most important, by far, is that the benefits depend critically on whether effective competition can be established and maintained in the industry. When competition exists, privatization has the potential to reduce costs and improve the quality of urban bus services; without competition, such reforms may bring little improvement and conceivably even a degradation in service or (as in Santiago) unwarranted increases in fares and excess capacity as route associations abuse their monopoly positions.

Interestingly, the prospects for effective competition among private bus operators seem greater in developing than developed countries. The striking feature of bus transit in the major cities of developing countries is the enormous number of small private operators, at least where local governments have not severely restricted entry. Minibus services are often provided by independent operators with one or at most a few vehicles each. Even standard-sized bus services are often offered by a half dozen or more operators each with small fleets of fifty or fewer vehicles. Competition is often further enhanced because the minibus and standard bus operators not only compete with one another, but with shared-ride taxis, jitneys, or motorized tricycles as well. By contrast, in Britain, with the most openly competitive bus transit of all developed countries and several years after deregulation and privatization, most metropolitan areas are dominated by one to three large bus firms, with smaller firms rarely accounting for more than 10 percent of the patronage.

Although the competitive prospects seem greater in developing countries, government intervention may still be needed to maintain competition, although it typically takes a different and less urgent form than in the industrialized countries. In industrialized countries when pro-competitive policies are pursued, the need is for vigilance against collusion among the few dominant firms usually found in each metropolitan area or predatory behavior undertaken by large firms against their smaller rivals. In developing countries the greater potential for anticompetitive behavior resides in the practices and procedures of route associations. Although these associations offer important benefits in coordinating schedules and reducing unsafe driving by small independent operators, the route associations also can, as Santiago illustrates, limit competition by restricting entry or encouraging higher fares.

Chapter 4. The British Experiment

PRIVATE BUS COMPANIES were common in Great Britain, the United States, and other developed countries in the first half of this century. As incomes and auto ownership began to grow and bus ridership fell, however, the developed countries started subsidizing their local bus services, and most local bus companies became publicly owned. Unfortunately, public ownership and growing public subsidies, often amounting to more than half the total cost of bus service, were not very successful in arresting the decline in bus ridership.[1] The cycle of bus system development and decay outlined in chapter 2 thus has been as common in the industrialized countries as in the developing world. The only big difference between the two areas is that the industrialized countries have been less likely to recycle from public back to private ownership since, being richer, they seem better able to finance the deficits of their public systems. Their efforts at bus privatization have usually been limited to greater contracting out of different operating functions (such as maintenance) or, occasionally, contracting out or franchising a few routes. The British, however, stand out as the principal exception to this general rule.[2]

The Scope of Britain's Reform

In one of the most dramatic and ambitious efforts ever undertaken to privatize local public services, the British Transport Act of 1985 ordered the deregulation and privatization of local bus services throughout Brit-

1. F. V. Webster and others, *Changing Patterns of Urban Travel*, report prepared by the Transport and Road Research Laboratory for the European Conference of Ministers of Transport (Paris: Organization for Economic Cooperation and Development, 1985); and John R. Meyer and José A. Gómez-Ibáñez, *Autos, Transit and Cities* (Harvard University Press, 1981).

2. New Zealand in June 1991 was the only other developed country to follow the British lead. Fiona Knight, "Structural Reform in New Zealand's Passenger Transport Industry," Manager for Policy and Planning, Transit New Zealand, March 1991.

ain, exempting only the Greater London metropolitan area.[3] The 1985 act, officially implemented on October 26, 1986, incorporated three key changes.[4] First, government controls over entry into the local bus industry were greatly relaxed so that public or private bus companies could offer virtually any bus service they deemed profitable by simply giving local authorities forty-two days' notice.[5] Second, the publicly owned bus companies that had dominated local bus service were reorganized as separate for-profit corporations. (Many of these companies were subsequently sold to the private sector, often through management or labor buyouts, while those that remained publicly owned could no longer receive direct government subsidies.) Third, local authorities could supplement the profitable or commercial routes by subsidizing additional services that they felt were warranted by social concern, but these supplementary services had to be secured through competitive bidding among the operators, with the lowest bidder normally awarded the service contract.[6] In short, Britain privatized and deregulated the local bus industry while still preserving, through competitive contracting, the possibility of subsidizing "socially worthwhile" but unprofitable services.

3. Ostensibly, Greater London was left out because, the year before the 1985 Transport Act, the national government had ordered the abolition of the Greater London Council (the metropolitan government of London) and the restructuring of London Transport (the public bus and subway operator). These reforms were thought to be change enough, at least for the time being. The government has required London Transport to contract out for a significant share of its bus services, however, and has announced its intention to deregulate London's local bus service in the early 1990s.

4. The analysis in this chapter is based primarily on José A. Gómez-Ibáñez and John R. Meyer, *Deregulating and Privatizing Urban Bus Services: Lessons from Britain*, report prepared for the Department of Transportation, Urban Mass Transportation Administration, January 1989. Other excellent sources include Laurie Pickup and others, *Bus Deregulation in the Metropolitan Areas*, Oxford Studies in Transport (Brookfield, Vt.: Avebury, 1991); and the many papers of Peter R. White including "Three Years' Experience of Bus Service Deregulation in Britain," paper presented at the International Conference on Privatization and Deregulation in Passenger Transportation, Tampere, Finland, June 1991; and R. P. Turner and Peter R. White, "Overall Impacts of Bus Deregulation in Britain, *Transportation Planning and Technology*, vol. 15 (1991), pp. 203–29.

5. In the past, authorities could reject applications to operate services they thought not needed or not in the public interest, but under the new law the only grounds for refusal are serious safety or traffic congestion problems.

6. The one exception to this rule is government reimbursements for concessionary fares (discounts from regular fares that are offered in many areas to the elderly and children). Most local authorities required some concessionary fares before deregulation and can continue to do so, even on commercial services, provided that they reimburse the operators for any revenues or profit lost because of the difference between concessionary and regular fares.

These reforms were applied to every local government across the country (excepting only the Greater London area), moreover, which allows comparison of their effectiveness in different circumstances. Great Britain outside of London divides into seven metropolitan counties and over forty shire counties (either rural or with only small cities). Of particular interest are the seven large British metropolitan areas outside London where U.K. local bus ridership is concentrated (table 4-1).[7] Although shire counties also include several medium-sized cities, much of the shire bus service is in rural areas; furthermore, the shock of deregulation, subsidy reduction, and privatization was less pronounced in the shires, since, before 1986, 90 percent of shire bus service costs were typically paid out of the farebox. By contrast, in the seven British metropolitan counties outside London, farebox revenues before deregulation covered only about 55 percent to 60 percent of costs.

Before their implementation, the British debated extensively and intensively about the probable effects of the proposed reforms.[8] Those in favor claimed that the reforms would create a market sufficiently competitive that any monopoly profits would be limited, if not eliminated. They also argued that these competitive pressures would force the industry to improve productivity and reduce costs and to devise improved and more attractive services. Largely because of the likelihood of reduced costs and greater competition, some proponents also predicted reduced fares and higher bus ridership.

Those opposed expected competition not to develop or to be wasteful if it did. They contended that many communities would end up dominated by entrenched local bus monopolies, which, being unresponsive to users and no longer inhibited by government oversight, would raise fares and cut services. If competition did develop, they said, it would be concentrated on a few profitable routes where competing firms would offer wasteful duplicative services, so that costs would escalate and productivity decline. Services would deteriorate on the unprofitable

7. County government was abolished in the seven metropolitan areas at about the same time as the bus reforms. To avoid confusion, we refer to those metropolitan areas as counties throughout, although, strictly speaking, they were counties only in the years before the reforms.

8. For example, U.K. Department of Transport, *Buses*, Cmnd 9300 (London: Her Majesty's Stationery Office, 1984); K. M. Gwilliam, C. A. Nash, and P. J. Mackie, "Deregulating the Bus Industry in Britain: (B) The Case Against," *Transport Reviews*, vol. 5 (April–June 1985), pp. 99–103; and Michael E. Beesley and Stephen Glaister, "Deregulating the Bus Industry in Britain: (C) Response," *Transport Reviews*, vol. 5 (April–June 1985), pp. 133–42.

TABLE 4-1. Characteristics of the Seven British Metropolitan Counties

County	Principal city	1985 population (thousands)[a]	Land area (square miles)	Density in persons per square mile	Cars owned per thousand persons (1984)[b]
Manchester	Manchester	2,583	497	5,197	241
Merseyside	Liverpool	1,481	252	5,877	215
South Yorkshire	Sheffield	1,303	602	2,163	240
Strathclyde	Glasgow	2,383[b]	2,185	1,091	191
Tyne and Wear	Newcastle	1,140	208	5,481	201
West Midlands	Birmingham	2,642	347	7,613	295
West Yorkshire	Leeds	2,053	787	2,609	248

a. U.K. Central Statistical Office, *Annual Abstract of Statistics 1987* (London: Her Majesty's Stationery Office, 1987).
b. Data for 1984 from Mervyn Jones, "Transport Policies and Practices in Britain's Conurbations, 1968–1986," paper prepared at the Transport Studies Unit, Oxford University, 1986.

routes, they predicted, especially in low-density areas and during off-peak hours, generating a disintegration in the system of coordinated services operated under the previous regime. As a result, bus ridership would decline and auto use and traffic congestion would increase.

Competition

A key dispute in the policy debate was whether the market would prove competitive. Within a year of the implementation of reforms only 3 percent of all the local bus mileage outside London was on routes served by more than one bus company; hence, most British bus passengers still had no choice of bus companies for any particular trip. Economists have long recognized, however, that actual competition is not essential if the markets are "contestable": that is, if incumbent firms believe that a challenger could easily enter their market, establishing a threat of entry sufficient to discourage an incumbent from raising prices or otherwise enjoying the fruits of apparent monopoly.[9] Significantly, almost every British county has seen entry in at least part of the local route network since deregulation, and many counties have seen nearly constant warfare between rival bus companies, though usually over only a small share of the routes at one time. Equally important, most bus companies have taken defensive measures to cut their costs or improve the competitive position

9. For a discussion of contestability see, for example, William J. Baumol, John C. Panzer, and Robert D. Willig, *Contestable Markets and the Theory of Industry Structure* (Harcourt Brace Jovanovich, 1982).

of their routes, even in areas where they have not been attacked. These measures suggest that the companies regard the threat of entry as credible.

Changes in Market Structure and Shares

Before deregulation the local bus industry included four principal types of bus operators outside London. Each type followed a somewhat different pattern in adapting to the 1985 legislation.

Before the reforms, two large nationalized companies, the National Bus Company (NBC) in England and Wales and the Scottish Bus Group (SBG) in Scotland, provided more than 50 percent of the local bus miles and carried almost 40 percent of local bus passengers outside of London, as shown in table 4-2. The NBC and the SBG were formed in 1969 by combining already nationalized bus companies with several private companies purchased by the national government that year. The NBC and SBG subsidiaries operated in both metropolitan and shire counties but were especially dominant in rural areas and small towns. The 1985 act ordered the seventy NBC subsidiaries sold separately to the private sector, and this sale was accomplished by early 1988 (netting the government more than 325 million pounds, substantially more than most observers expected). The ten SBG subsidiaries were not required to be sold, but the Scottish office began exploring the possibilities in the late 1980s.

Approximately 25 percent of the prereform bus miles and 40 percent of the passengers were attributable to seven passenger transport companies (PTCs), one each in the seven major metropolitan areas. Each PTC was owned by a passenger transport executive (PTE), which was in turn controlled by the metropolitan county council.[10] The PTE was responsible for bus service in the metropolitan area, most of which it provided directly through its PTC; in some metropolitan areas the PTE supported additional local bus services provided by NBC or SBG subsidiaries. The 1985 act required that each PTE "corporatize" its PTC (that is, spin it off as a separate, for-profit company that could no longer receive direct subsidies). The PTCs did not have to be sold off to the private sector, although two of the seven had been by 1991.

Forty-four of the larger cities in the shire counties had a municipal bus company, collectively accounting for about 12 percent of the prereform

10. The PTEs were called passenger transport authorities before the reforms; the term PTE is used throughout for the sake of simplicity. Also, at about the same time as the reforms the national government abolished the elected metropolitan county councils; the bus PTEs were then controlled by municipal and other local governments in their jurisdictions.

TABLE 4-2. Percentage of Vehicle-Miles Operated and Passengers Carried by Different Types of Operators before and after Deregulation, Excluding London[a]

	Public or ex-public operators				
Item	*Ex-national bus company*	*Scottish bus group*	*Passenger transport companies*	*Municipal*	*Private independents*
Vehicle-miles operated (fiscal years)					
Before					
1985-86	43.1	9.3	26.0	12.1	5.5
Early 1986–87[b]	44.5	9.2	24.9	11.9	5.5
After					
Late 1986–87[b]	43.2	9.0	22.2	11.5	14.0
November 1987	42.6	8.6	21.7	12.1	15.0
1987–88	41.9	8.2	22.3	11.5	16.0
Passengers carried (fiscal years)					
Before					
1985–86	31.2	6.8	40.5	17.3	3.9
During					
1986–87[b]	32.3	7.0	37.8	17.8	5.2
After					
1987–88	32.5	6.7	35.9	16.8	8.1

a. Calculated from data in U.K. Department of Transport, *Bus and Coach Statistics, 1987/88* (London: Her Majesty's Stationery Office, 1989).
b. Fiscal year 1986–87 includes seven months before and five months after deregulation day.

mileage and 17 percent of the passengers. Under the 1985 act the municipal companies could remain publicly owned but, like the PTCs, had to be established as arm's-length and for-profit companies. Again, many have been privatized.

Finally, a number of private independent bus companies existed before deregulation and accounted for approximately 5 percent of prereform service. Most were very small and engaged in commercial services in rural counties or subsidized services in metropolitan counties. Some of these have greatly expanded since the reforms, by simply extending their routes or by acquiring some of the previously public companies. Specifically, the private independents' share of vehicle miles increased from 5.5 percent immediately before deregulation to 15.0 percent only a year later, with much of the loss suffered by the metropolitan PTCs (table 4-2); the independents' share of local bus passengers increased from 3.9 percent in the fiscal year before the reforms to 8.1 percent in the first full fiscal year after deregulation (which started five months after deregulation day).

Effective competition has varied significantly among counties, as suggested by the shifts in market shares in the metropolitan counties shown in table 4-3. In two of the seven metropolitan counties, Strathclyde and Manchester, competition has been extremely intense. In Strathclyde before deregulation, the PTC's buses dominated Glasgow city service. After deregulation, three SBG subsidiaries that previously operated mainly in the suburbs extended their routes into Glasgow and the newly independent and corporatized PTC retaliated by extending its routes into the suburbs. Market shares apparently did not change greatly, since the SBG subsidiaries and the PTC were fairly evenly matched. In Manchester, the PTC's own buses accounted for 97 percent of the mileage before deregulation, but an outside bus company began a major minibus operation in the southern part of the territory and at one point accounted for over a quarter of the county's mileage.

By contrast, in one of the seven metropolitan areas, West Midlands, there has been relatively little effective competition. The PTC, which accounted for 96 percent of all mileage before deregulation, has never really faced any serious or sustained challenges. In part, this is thought to be a consequence of fares initially being relatively low in the West Midlands; the West Midlands PTC is also generally regarded as being relatively efficient and aggressive.

In the four remaining metropolitan counties (Merseyside, Tyne and Wear, South Yorkshire, and West Yorkshire), competition has been at an intermediate level. Before deregulation, those counties were usually served by several NBC subsidiaries as well as a PTC. Unlike the companies in Strathclyde, however, the PTC and the ex-NBC subsidiaries in these four counties apparently reached an implicit understanding that they would not attack one another's traditional territories. Most of the competition has been from private independent firms, which increased their share of mileage from less than 2 percent to 6 or 7 percent in the first year alone.

Strategies of the New Entrants and the Old Guard

The new entrant challengers and the large incumbent old guard seem to have been fairly evenly matched in most counties, partly because both types of competitors have developed relatively effective strategies. A key element in the competitive strategy of the new challengers has been to keep unit costs very low, partly through paying lower wage rates but also through lower overheads and more flexible work rules. Most new en-

TABLE 4-3. Shares of Bus Mileage by Type of Operator, before and after Deregulation, for Selected Metropolitan Counties

Level of competition, metropolitan county, and date	*PTC*	*Ex-NBC/SBG*	*Private independents*
High competition			
Strathclyde			
Before			
November 1985	33.2	54.0	12.8
After			
November 1986	26.5	57.8	15.7
August 1987	29.9	53.6	16.5
November 1987	30.8	52.8	16.4
Manchester			
Before			
November 1985	97.0	4.0	0.6
After			
November 1986	85.1	5.7	9.2
November 1987	67.0	9.0	26.0
Intermediate competition			
Merseyside			
Before			
November 1985	76.0	22.0	2.0
After			
November 1986	65.2	30.9	3.9
April 1987	69.9	22.9	7.2
South Yorkshire	n.a.	n.a.	n.a.
Tyne and Wear			
Before			
November 1985	39.5	59.1	1.4
After			
January 1988	37.5	56.0	6.5
West Yorkshire			
Before			
November 1985	55.0	43.0	2.0
After			
November 1986	51.0	43.0	6.0
November 1987	52.0	41.0	7.0
Low competition			
West Midlands			
Before			
October 1986	95.9	4.1	0
After			
November 1986	92.8	6.7	0.5

SOURCES: Calculated from various unpublished and published reports of the Metropolitan Passenger Transport Executives and the U.K. Transport and Road Research Laboratory; see José A. Gómez-Ibáñez and John R. Meyer, *Deregulating and Privatizing Urban Bus Services: Lessons from Britain*, report prepared for the Department of Transportation, Urban Mass Transportation Administration (Washington, January 1989), p. 30.
n.a. Not available.

trants chose not to exploit this cost advantage by charging lower fares, mainly because of a perception that fares would have to be much lower to induce a passenger to wait at the stop for another company's bus. Instead, lower costs helped them avoid retaliation from the larger incumbent firms by concentrating on commercial routes that are only marginally profitable for the higher-cost old guard firms or by underbidding the old guard for those services that continued to be subsidized.

While most of the new entrants in the first year of deregulation were small firms, a new type emerged beginning in the second year: the subsidiary of a holding company with bus operations in several counties. About a dozen such holding companies evolved. Most started as a private independent firm or as one of the first ex-NBC subsidiaries to be sold off. They then expanded to other territories, usually distant from their original bases, by buying and reinvigorating an existing bus company or by starting a whole new subsidiary. These holding companies combined the very low costs of new entrants with the flexibility and spreading of risk provided by diversified holdings in several areas. Vehicles and management talent can also be shifted from one subsidiary to another as market conditions dictated. Furthermore, with a holding company, more opportunities can be offered to managers, and treasury, finance, and other staff overhead functions can be spread over more units of output.

The large incumbents responded by improving productivity, cutting work rules, and, in many cities after a year or two, establishing a two-tier wage scale with lower rates for newly hired drivers. The large firms also developed interesting strategies to turn their size and incumbency into an advantage. While there are apparently few or no economies of scale in local bus services,[11] size does at least offer a better opportunity for cross subsidy (for example, from profits on segments that are not under attack). Indeed, most old guard firms attempted to deter entry by matching, or over matching, new services offered by major challengers, a practice that became known as "route swamping" in the British industry.

Some large incumbents also used travel cards—a monthly, weekly, or daily pass that offers unlimited travel for a fixed price—as a means of turning size into a competitive advantage. Before deregulation, the county authorities that were subsidizing bus service often negotiated travel cards

11. See, for example, N. Lee and I. Steedman, "Economies of Scale in Bus Transport," *Journal of Transport Economics and Policy*, vol. 4 (January 1970), pp. 15–28; and K. Button and K. O'Donnell, "An Examination of the Cost Structures Associated with Providing Urban Bus Services in Britain," *Scottish Journal of Political Economy*, vol. 32 (February 1985), pp. 67–81.

that could be used on any local company's buses. Many large firms dropped out of these countywide travel card schemes and set up their own competing company card; the reasoning, as with U.S. airlines' frequent flyer programs, was that a card from a company with a large network would be much more attractive than a card from a company with a small network.

Although size offers the large incumbents important advantages, their strategies are not without risks and limitations. Large incumbents cannot engage in extensive cross subsidies without making their profitable routes more tempting targets for competitors and thus inviting further attacks. Travel cards and other network premiums can attract passengers to operators with large networks but, as the U.S. airlines discovered with their frequent flyer programs, probably not without incurring substantial financial costs.

Large incumbents also have competitive weaknesses that offset some of their advantages. They are unlikely to ever lower their costs to the levels possible for a new entrant who can start out with low overhead, few workers with much seniority, and a fresh labor agreement. A large incumbent is especially disadvantaged when local authorities contract for subsidized services, both because success in bidding depends heavily on low unit costs and because a new entrant often enjoys at least a temporary bidding advantage if the subsidized service attracts some passengers from the incumbent's routes. Finally, the large incumbents also may face a growing challenge from holding companies that enjoy some of the incumbent's advantages of large size (particularly for cross subsidy) without all of the disadvantages (for example, higher costs).

The Government's Role

The framers of the 1985 Transport Act feared that local authorities might undermine rather than promote competition. The act therefore specifically enjoined local authorities from designing subsidized or contract services that compete with commercial services.

Despite the national government's fears, local authorities have usually encouraged competition. A principal impetus for these procompetitive policies was the relatively low number of bids on the initial rounds of subsidized service contracts solicited in 1986; the paucity of bids left local authorities fearful that their subsidy budgets would eventually prove inadequate to maintain all the services they thought socially necessary. In most metropolitan counties the local authorities elicited more bids on subsequent rounds by actively recruiting small bus operators, simplifying bidding procedures, and reducing the risk for potential suppliers.[12] One remaining problem has been that the incumbents often retaliate when they lose a bid for subsidized service by registering that route for commercial (unsubsidized) service. Under national government regulations banning inhibition of competition, such a measure requires local authorities to cancel the subsidized service contract.[13]

The national government in the long run may need to play a greater role in maintaining competition, particularly in enforcing antitrust laws to keep some of the anticompetitive strategies of the large incumbents and the emerging holding companies in check. In a series of cases since deregulation, national antitrust authorities have acted to stop incumbents from restricting new entrants' access to city bus stations, to discourage predatory retaliation by incumbents (such as dropping prices below variable costs or unprofitable matching of services or timetables), and to order some mergers of neighboring companies to be dissolved. Some British observers argue that the government has overemphasized the dissolution of anticompetitive mergers, however, and not done enough to prevent predatory retaliation. Britain's antitrust statues are relatively weak (at least by U.S. standards), moreover, in that sanctions are modest and no practices in themselves are illegal so each case must be decided on its merits.[14]

Changes in Service, Costs, Fares, and Ridership

The changes in Britain's subsidies, service, fares, and ridership in the first two fiscal years of regulation are summarized in table 4-4.

12. William J. Tyson, "Subsidizing Bus Services by Competitive Tender: Initial Experience in U.K. Metropolitan Areas," report to London Transport International, Manchester, June 18, 1987; and William J. Tyson, "A Review of the First Year of Bus Deregulation," report to the Association of Metropolitan Authorities and the Passenger Transport Executive Group, London, April 1988.

13. Local authorities have experimented with various strategies for encouraging bidding in this environment, such as maintaining the subsidized service when the overlap between commercial and subsidized services is modest or writing flexible contracts that allow them to redeploy the low bidders' buses on another subsidized route if the incumbent retaliates with a commercial service.

14. See Gómez-Ibáñez and Meyer, *Deregulation and Privatizing Urban Bus Services*, pp. 48–51; and Jonathan R. Preston, "Competition Policy and the British Bus Industry: The Case of Mergers," and J.S. Dodgson, "Predatory Behavior in the Passenger Transportation Industry," both papers presented at the International Conference on Privatization and Deregulation in Passenger Transportation, Tampere, Finland, June 1991.

Subsidy Cuts and Other Complicating Factors

Isolating the effects of privatization on service, fares, and ridership is complicated by the fact that government subsidies were cut simultaneously. The cuts were especially harsh in the metropolitan counties where, over a two-year period commencing seven months before deregulation day, real subsidies to support bus services were cut by about 23 percent.[15] Two of the metropolitan counties that had a policy of holding fares down to nominal levels (often only 5 pence) were forced to cut subsidies by 40 percent to 47 percent (see table 4-4). In the shire counties, by contrast, bus subsidies were cut by about 6 percent during those same two years.

A second complicating consideration is that trends exogenous to the industry, such as inflation and economic growth, would have caused some changes in bus costs, fares, and ridership even in the absence of the 1985 act. For example, even without the reforms, industry costs and fares probably would have increased at the rate of general price inflation, and thus only changes in real fares and costs (that is, above and beyond inflation) are pertinent for analyzing the impact of the 1985 act. The reforms were also implemented during a period of national economic recovery and declining real gasoline prices; statistical analysis of past years suggests that these trends would have caused losses in bus ridership of 2 percent to 3 percent a year, even in the absence of bus fare increases or service cuts.[16]

A final complicating factor was the confusion, some would even call it the chaos, experienced during the abrupt transition to the new regime. All companies intending to offer commercial services on the first day of deregulation (October 26, 1986) had to register those services by March 1986 to allow local authorities seven months to identify gaps in the commercial service and to design and award contracts for subsidized services. The incumbent public bus companies had to plan for their new operations and to continue to provide services under the old system until midnight, October 25, 1986. It was difficult for them to anticipate the number of subsidized service contracts they would be awarded, shed staff and vehicles, and train drivers and dispatchers for the new routes in an orderly fashion. On deregulation day, there were many stories of passen-

15. Gómez-Ibáñez and Meyer, *Deregulating and Privatizing Urban Bus Services*, p. 137.

16. Gómez-Ibáñez and Meyer, *Deregulating and Privatizing Urban Bus Services*, pp. 231–41.

TABLE 4-4. Changes in Bus Subsidies, Service, Fares, and Ridership after Deregulation

Percent unless otherwise noted

Item[a]	*Bus subsidies (constant pounds)*[b]	*Bus miles of service*	*Bus fares (constant pounds)*	*Bus ridership*	*Evaluation of: Deregulation transition*	*Evaluation of: Level of competition*
Six English metropolitan counties (fiscal years)						
One year (1985/86–1986/87)[c]	–12.6	–3.0	23.9	–12.5	...	...
Two years (1985/86–1987/88)[c]	–27.6	7.5[a]	28.5[a]	–16.2[a]	...	...
Other counties outside London shires plus Strathclyde						
One year (1985/86–1986/87)[c]	–1.2	7.7	1.1	–3.0	...	...
Two years (1985/86–1987/88)[c]	–6.3	16.5	1.0	–3.8	...	...
Seven metropolitan counties (fiscal 1985–86–1987–88)						
Manchester	–5.3	14.6	7.2	–9.3	Poor	Strong
Merseyside	–47.1	–0.2	39.1	–32.3	Poor	Average
South Yorkshire	–39.8	–7.7	177.1	–29.2	Average	Average
Strathclyde	n.a.	13.7	0.5	–0.3	Average	Strong
Tyne and Wear	n.a.	–3.5	35.5	–11.5	Average	Average
West Midlands	4.5	–1.8	4.9	–13.4	Average	Low
West Yorkshire	–10.2	5.3	–4.9	–2.4	Average	Average
All seven counties	–23.4	3.3	35.0[a]	–14.1[c]	...	...

n.a. Not available.

a. The regional statistics for six English metropolitan counties are calculated from U.K. Department of Transport, *Bus and Coach Statistics for Great Britain, 1988/89* (Her Majesty's Stationery Office, 1989), pp. 19–22, 24, 35–36. The statistics for individual metropolitan counties are calculated from various published and unpublished reports of the Metropolitan Passenger Transport Executives and the U.K. Transport and Road Research Laboratory (see Gómez-Ibáñez and Meyer, *Deregulating and Privatizing Urban Bus Services*, pp. 92, 134). The statistics from these two sources are not entirely consistent with each other, although it is difficult to make exact comparisons because the U.K. Department of Transport does not report statistics for the seven metropolitan counties individually or as a group (only for the six English metropolitan counties as a group). The U.K. Department of Transport suggests, for example, that fare increases were smaller and service increases greater than the individual metropolitan county statistics show.

b. Includes general support, concessionary fare support, and rural bus grants, but excludes the fuel duty rebate.

c. Fiscal year 1985–86 is the last full fiscal year before deregulation. Fiscal year 1986–87, which began April 1, 1986, includes seven months leading up to and five months after deregulation day. Fiscal year 1987–88 is the first full fiscal year after deregulation.

gers confused by route changes, buses not operating as scheduled, and even drivers asking passengers for directions.

In some areas, local authorities unintentionally added to the transition problems. For example, Manchester authorities contracted for relatively few subsidized services to start on deregulation day out of a misplaced (if understandable) fear that their subsidy budget might be quickly exhausted; it took up to six months for many services to be restored. In

Merseyside local authorities decided to use a new system for numbering routes for their subsidized services and failed to publicize these changes until well after deregulation day, thus contributing greatly to passengers' confusion. Some industry observers worry that, in areas like Manchester and Merseyside, where transition problems were unusually severe, service instability may have permanently driven away many bus riders.

Because all of Great Britain except metropolitan London was privatized and deregulated, moreover, there is no obvious benchmark or "control" area against which to measure the effects of the 1985 act. Some analysts have suggested that Greater London should be the benchmark, but London's economy was expanding and traffic congestion increasing unusually rapidly at the time. London's public transport companies, although neither deregulated nor privatized, were also subject to new competitive pressures because the national government required that they do additional contracting out during this period.

Service

Measured in total bus mileage operated, service has increased since the reforms. An astonishing 80 percent of the prederegulation mileage nationwide was registered for commercial (that is, unsubsidized) service on deregulation day, and the level grew subsequently.[17] As a result, local authorities, despite subsidy cuts, contracted for enough subsidized service to bring total mileage above prederegulation levels in most areas (see the second column of table 4-4). Specifically, in the metropolitan counties total mileage was at least 3.3 percent higher in the first full fiscal year after deregulation (1987–88) than it had been in the last full fiscal year before deregulation (1985–86). In the English shire counties the trend was even more favorable, with mileage rising by about 16.5 percent over the same period.

Deregulation also encouraged service innovations, the most dramatic of which was the spread of twelve- to twenty-five-seat minibuses to replace or augment conventional services in forty- to eighty-seat single- or double-deck buses. Minibuses were actually pioneered by a public bus company before deregulation, but they spread rapidly under the new

17. By fiscal years 1987–88 and 1988–89, for example, commercial service amounted to 95.1 percent and 98.9 percent, respectively, of the mileage operated in the last fiscal year before deregulation in all of Great Britain outside London; U.K. Department of Transport, *Bus and Coach Statistics, Great Britain: 1988/89* (London: Her Majesty's Stationery Office, 1989), p. 21.

competitive pressures.[18] Just three years after deregulation, small vehicles had grown from a few hundred to 7,000 of the 30,000 buses providing local service outside of London.[19] Minibuses have lower operating costs per vehicle-mile, not only because the drivers are usually paid less than big bus drivers but also because the vehicles are simpler, more easily maintained, and operate at higher average speeds (since each vehicle stops less often to pick up passengers and has greater maneuverability in traffic).[20]

The minibuses also offered service advantages to customers in the form of higher frequencies or more direct routings, as well as higher speeds. Minibuses, owing to their smaller size, also can venture further into residential housing developments and other areas that tend to have traffic volumes or roads too limited to accommodate big buses. The small size of the minibus also means that a route can have several branches to serve different neighborhoods at its outer end and still maintain reasonable headways on both the trunk route and each branch. The mini's maneuverability and size have also led some operators to offer "hail and ride" services in which the minibus will stop at any point on the route to pick up or discharge passengers. Improved route coverage and hail and ride reduce walking distances and make bus travel more accessible, especially for elderly and disabled passengers. An improved sense of safety owing to closer driver proximity and the greater likelihood of traveling with local neighbors of a similar "class" (because of the smaller bus size) have also been cited as factors making minibuses more attractive.

These service advantages have translated into some dramatic increases in patronage. The most successful applications have generally been in places where low-frequency conventional bus service was replaced by high-frequency minibus service. In one city, for example, ridership increased by 180 percent when double-deck buses every twenty to thirty minutes were replaced by minibuses every three to five minutes.[21] Signifi-

18. The most widely credited pioneer is Devon General, an NBC subsidiary that began implementing minibuses in 1984; see Harry Blundred, "Buses: Public Service or Private Profit?" *Transport Planning and Technology*, vol. 15 (1991), pp. 107–14.

19. White, "Three Year's Experience," p. 8; and Peter R. White, "An Overview of Urban Minibus Operations," paper presented at the symposium Experience of Urban Minibus Operation in Britain, City University, London, May 1988.

20. Minibus drivers are generally paid about 20 percent less than conventional bus drivers; the difference in wage rates accounts for only about two-fifths of the savings in operating cost per vehicle mile, however, with the balance attributable to reductions in maintenance and other expenses; see Gómez-Ibáñez and Meyer, *Deregulating and Privatizing Urban Bus Services*, pp. 118–19.

21. Blundred, "Buses," p. 109; see also White, "Three Years' Experience," p. 11.

cant increases in frequency are profitable even if they are expected to generate only modest increases in ridership, moreover, since British operators found that minibuses cost only about 60 percent as much as a conventional bus to operate per vehicle-mile.[22] In places where conventional bus frequencies were already high, however, the introduction of high-frequency minibuses did not have as great an effect on patronage.

The introduction of minibuses reversed a long-standing trend in Britain of reducing unit costs, usually at the expense of less frequent service, by buying big double-deck and articulated buses. Indeed, most bus industry experts thought that minibuses were a technology more suited for the developing world, where wages were lower.[23] Whether these small vehicles will keep growing in popularity is a hotly debated topic within the British industry.

Deregulation also may have stimulated local British authorities to be more innovative in the services they subsidize. Now that contracts for subsidized services are bid competitively, local authorities know how much individual services cost. This new information has stimulated some authorities to abandon subsidized services whose costs seem to exceed the perceived social benefits and to experiment with other alternatives.

As opponents of the reforms predicted, not all areas and passengers have experienced service improvements. For example, relatively few off-peak and low-density services were registered on a commercial basis. Although local authorities contracted to restore most of the services that were not commercially registered, in virtually every county there were some places and times of day with less service two years after deregulation than before.[24]

Nevertheless, service as a whole probably has improved. The new services are the ones that private firms think passengers will value enough to pay for, while the lost services are mostly ones cut by the newly privatized firms and not restored by public authorities, mainly because they are the least valued by customers. It must also be remembered that

22. Gómez-Ibáñez and Meyer, *Deregulating and Privatizing Urban Bus Services*, pp. 117–18.

23. A notable exception to this conventional wisdom was Herbert Mohring; see Mohring, "Minibuses in Urban Transportation," *Journal of Urban Economics*, vol. 14 (November 1983), pp. 293–317.

24. For example, see R. J. Balcombe, J. M. Hopkin, and K. E. Perrett, *Bus Deregulation in Great Britain: A Review of the First Year*, Research Report 161 (Crowthorne, England: U.K. Department of Transport, Transport and Road and Research Laboratory, 1988).

the authorities' subsidy budgets were sufficient to restore total mileage to prederegulation levels in most areas, so deletion of a particular previous service suggests that it was little valued.

The pattern of service changes among the seven metropolitan counties, as shown in table 4-4, indicates that competition was especially important in encouraging service improvements and offsetting the effects of subsidy cuts. The two metropolitan counties where competition has been most intense, Manchester and Strathclyde, have the largest mileage increases, while the county with the least competition, West Midlands, has a mileage loss, even though it is also the only metropolitan county where bus subsidies actually increased.

Costs, Productivity, and Labor

The newly privatized bus companies regard cost data as proprietary and, even where cost data are available, before and after cost comparisons are complicated because the service supplied almost never stayed the same. Several of the previously least-efficient PTCs report reducing their costs per vehicle-mile by about 25 percent, from about 2.0 British pounds before to 1.5 pounds per double-deck bus mile after deregulation. Some of the new entrants reported costs below 1.0 British pound per double-deck bus mile in 1987, however, which suggests even larger cost reductions.[25]

Perhaps the best estimates of the likely cost savings come from London, which was not deregulated but where contracting out is used extensively. Careful cost comparisons by London Transport before and after contracting suggest a savings, above and beyond London Transport's contract monitoring costs, of about 20 percent, with no reduction in service quality.[26] Moreover, there are signs that contracting has stimulated the public bus companies in London to reduce their costs as well, for fear of encouraging authorities to contract for even more services.

Fares and Ridership

After 1986, fares increased faster than inflation, despite cost savings, rising by about 25 percent in real terms (that is, net of inflation) in the seven metropolitan counties during the fiscal year that includes deregulation day, and by another 10 percent in the subsequent fiscal year. Rider-

25. Gómez-Ibáñez and Meyer, *Deregulating and Privatizing Urban Bus Services*, pp. 34, 37–38, 112–14.

26. David Bayliss, "Bus Service Tendering in London," paper presented at the Congress of the International Union of Public Transport, Lausanne, Switzerland, May 1987.

ship in the metropolitan counties fell by about 14 percent compared with prederegulation levels. This decline, however, is very much in line with what would have been predicted because of the fare increases and the established downward secular trend in transit patronage (even allowing for the positive effects on patronage of deregulation-induced service improvements). The record in the shire counties was a little better, with a 1 percent fare increase in the first two years, accompanied by a 4 percent ridership loss. This is again roughly consistent with what might be expected from historical trends in ridership plus the fare and service changes.

The pattern of fare increases and subsidy cuts across the seven metropolitan areas suggests the important connection between the two. Among the metropolitan counties shown in table 4-4, for example, the largest fare increases were in two counties, South Yorkshire and Merseyside, that for years had pursued policies of holding fares to nominal levels and thus were forced by the national government to accept the largest subsidy cuts.

Changes in Rail and Auto Use

The losses in bus ridership do not seem to have caused comparable increases in auto or rail use. Several metropolitan counties reported increases in commuter rail ridership after deregulation, but the rail gains were fairly small compared with the bus losses, perhaps because commuter rail is oriented to serving longer trips and thus is not often an easy substitute for bus service. Two of the metropolitan counties also have small subway (underground) systems (Tyne and Wear and Strathclyde), and in at least one case subway ridership came down because many new commercial bus routes ran express to the downtown after deregulation rather than terminating at outlying rail transit stops.[27]

Somewhat surprisingly there were no reports of significant increases in auto use attributable to deregulation. To the contrary, a survey of household travel patterns conducted periodically in Manchester reports that bus tripmaking declined from an average of 3.2 trips to 2.7 trips per person per week between spring 1986 and spring 1987 (about six months after deregulation, and just as bus competition in Manchester started to increase). At the same time, train travel increased from 0.19 to only 0.22 trips per person per week, and car and motorcycle trips went up from 8.8

27. Gómez-Ibáñez and Meyer, *Deregulating and Privatizing Urban Bus Services*, pp. 88–91.

to only 8.9 trips per week. Instead of shifting modes, bus users apparently just made fewer trips, as total trips by all modes declined from 15.7 trips to 15.2 trips per person per week.[28]

Isolating the Effects of the Reforms

Two simple calculations reported in table 4-5 suggest that the deregulation and privatization reforms actually softened the blows of subsidy cuts and adverse external trends, especially where competition was effective and transitional problems were not severe. The first calculation, shown in the first three columns of table 4-5, compares the actual changes in ridership in the seven metropolitan counties with the ridership changes that might have been expected from the fare and service changes implemented during deregulation and because of external trends. The trend effects are assumed to reduce ridership by 2.5 percent a year, a reasonable estimate from statistical analyses of past such effects.[29] Standard rules of thumb employed by the British bus industry (and widely accepted elsewhere, as in the United States) can be used to predict the effect of the fare and service changes: each percentage point (real) increase in fares should induce a 0.3 percent loss in ridership, while each percentage change up or down in bus miles brings a corresponding 0.4 percent change in ridership.

While the expected or predicted ridership loss was nearly identical to the actual loss (13.9 versus 14.1 percent) for all seven metropolitan counties as a whole, in three counties the expected loss was greater than the actual, in three it was less, and in one county expected and actual were about the same. In South Yorkshire, where the ridership loss was less than expected, the fare increase was so large, an average of 177 percent, that the usual industry rules of thumb about price elasticities probably do not apply. The three counties that fared worse than expected include the two where local authorities exacerbated transitional problems (Manchester and Merseyside) and the only county with little effective competition (West Midlands). Overall, where the bus reforms were well implemented and effective competition was established, less ridership was lost than might have otherwise been anticipated given the fare increases and service changes that occurred.

28. Harris Research Centre, *Greater Manchester Travel Diary, Spring 1987*, report prepared for the Greater Manchester Passenger Executive, August 1987.

29. Gómez-Ibáñez and Meyer, *Deregulating and Privatizing Urban Bus Services*, pp. 237–38.

TABLE 4-5. Expected and Actual Changes in Ridership and Expected Changes in Costs and Profitability for Metropolitan Bus Companies, Fiscal Years 1985–86, 1987–88[a]

	Percentage ridership changes			*Percentage revenue and cost changes*		
Region	*Expected*[b]	*Actual*	*Difference*[c]	*Actual change in total revenues (farebox and government subsidy)*[d]	*Hypothetical change in total costs if no productivity improvement*[e]	*Percent deficit (or profit) from revenue and cost changes*[f]
Manchester	−1.7	−9.3	−7.6	−3.8	14.6	18.4
Merseyside	−16.2	−32.3	−16.1	−30.9	−0.2	30.7
South Yorkshire	−56.7	−29.2	27.5	−9.3	−7.7	1.6
Strathclyde	−0.4	−0.3	0.1	n.a.	13.7	n.a.
Tyne and Wear	−16.4	−11.5	4.9	n.a.	−3.5	n.a.
West Midlands	−7.1	−13.4	−6.3	4.5	−1.8	(6.3)
West Yorkshire	−5.6	2.4	3.2	−10.2	5.3	15.5
All metropolitan counties	−13.9	−14.1	−0.2	−2.8	3.3	6.1

n.a. Not available.

a. Derived from sources described in José A. Gómez-Ibáñez and John R. Meyer, *Deregulating and Privatizing Urban Bus Services*, pp. 91–93, 132–38.

b. The forecast of patrons was derived by using a 0.4 elasticity to estimate service or bus mileage effects, a −0.3 elasticity to estimate fare effects, and assuming a −2.5 percent trend effect for the two years.

c. The patronage difference is equal to the actual patronage minus the forecast of patronage; therefore a negative difference indicates a greater decline in patronage under deregulation than would be predicted by fare, service, and trend effects alone.

d. Calculated using the actual changes in government and farebox revenues weighted by the 1985–86 percentages of revenues received from government, fares, and other sources.

e. Assumed in proportion to changes in bus miles reported in table 4-4.

f. Difference between the hypothesized change and the actual change.

These fare and service changes were also less severe than normally would have been needed to offset the government's subsidy cuts in most metropolitan counties, as demonstrated by the calculations in the last three columns of table 4-5. For all seven metropolitan areas as a whole, for example, actual bus company revenues (fares plus subsidies) declined by 2.8 percent while costs, absent any productivity improvements, would have increased by 3.3 percent, largely because of the increase in vehicle-miles of service. This combination of revenue declines and cost increases would have resulted in hypothetical bus company deficits (after subsidy) of 6.1 percent of total costs. The only metropolitan county where the bus companies would have been expected to profit from the revenue and service changes is West Midlands, which is also the only county where a local monopoly was unchallenged. What apparently saved most bus companies in the other counties was the productivity or unit cost reductions that were stimulated by the prospect, or reality, of competition. Without such cost reductions, the subsidy cuts would have required larger fare increases or smaller service increases.

Consistent with the projections in table 4-5, many of the large incumbents reported large losses in the first year of deregulation, but an increasing number of incumbents and many of the new entrants subsequently reported profits. Indeed, if companies cut unit costs by an average of 20 percent, as the London contracting experience suggests, the savings would more than offset the hypothetical company deficits reported in the last column of table 4-5 in all but one metropolitan area (Merseyside). Some companies may have made substantial profits even in the first year or two after the reforms. This group would probably include the leanest and the least competitively embattled of the new entrants and those old guard firms that have managed to cut costs or avoid competition. Profitability for the industry as a whole reportedly increased steadily in the first three years after deregulation, although some companies were clearly still in difficulty.[30]

Despite these favorable data, some analysts remain pessimistic in their assessments of the British experience. Peter White argues that the ridership losses in the metropolitan areas after the reforms were greater than should have been expected given the service mile and fare increases and the exogenous trends.[31] This result suggests, according to White, that passengers received little or no benefit from the large increases in bus miles operated in the metropolitan areas; deregulation and competition may have actually reduced the quality of service by stimulating frequent service changes, wasteful duplication of services, and the deterioration of countywide pass schemes. By contrast, London's ridership changes were more in line with, or better than, those expected from its service and fare changes according to White's calculations. This he attributes to the fact that London was subject only to contracting out and thus was spared the chaos and uncoordinated competition that accompanied more complete deregulation and privatization. White's results depend importantly on applying a much smaller exogenous trend in ridership loss than seems

30. White, "Three Years' Experience," pp. 7–9; and Gómez-Ibáñez and Meyer, *Deregulating and Privatizing Urban Bus Services*, pp. 135–38.

31. See White's numerous articles including Peter R. White, "A Welfare Assessment of Deregulation," *Journal of Transport Economics and Policy*, vol. 24 (September 1990), pp. 311–32; Turner and White, "Overall Impacts;" and White, "Three Year's Experience."

defensible given the historical record, however;[32] he also assumes that the same trend applies to both London and the other metropolitan areas, even though employment and population growth were generally stronger in London than elsewhere at the time.[33] This is not to deny that transitional problems occurred, or that competition sometimes means duplication of services that may be wasteful in the short term. Nevertheless, White's analysis seems unduly pessimistic. Moreover, contracting out, as we shall point out in the next chapter on the U.S. experience, may stimulate cost reductions but not necessarily the service innovations, such as minibuses, that have been such an important part of the British experience.

Winners, Losers, and Lessons

The clearest winners from the combined package of deregulation, privatization, and subsidy cuts are British taxpayers, who saw government expenditures for local transport in metropolitan counties cut by about a quarter in real terms in only two years. The owners of at least some of the bus companies also may have gained from deregulation, especially since the costs wrung from the industry seem to be larger than those required to offset the subsidy reductions and ridership losses.

Whether bus riders lost or gained is more difficult to determine. On the one hand, some riders clearly have lost, especially those who are no longer riding and those who still ride but have been asked to pay much higher fares; in two metropolitan counties where fares had not changed

32. White assumes an exogenous trend of 1.5 percent a year while we assume 2.5 percent. The source of White's trend is unclear but seems to be based on a straightline extrapolation of past trends. Our estimate is based on a multivariate statistical analysis of national ridership trends from 1964 to 1984. Depending on the exact specification of the equation used, this analysis suggests that bus ridership should have decreased by 3.6 to 4.7 percent in the first year after deregulation had fares and services remained constant. Much of the loss in ridership is attributable to the growth in real per capita incomes and the sharp decline in auto operating costs during that year. Thus even the 2.5 percent trend we assume is probably too conservative and understates the effects of the reforms; see Gómez-Ibáñez and Meyer, *Deregulating and Privatizing Urban Bus Services*, pp. 237–38.

33. Comparative statistics are not readily available, but in the mid-1980s the economy of London was booming while many of the seven metropolitan areas were experiencing much weaker growth. Moreover, London's ridership growth is also thought to have been stimulated by still-growing popularity of the prepaid travel card, first introduced in 1982. Travel cards may have more attraction in London than elsewhere because of the high levels of congestion, even in off-peak hours, and London's extensive underground subway system. Indeed, London's transit ridership has grown twice as fast in the off-peak time as in the peak time since travel cards were introduced.

for many years, South Yorkshire and Merseyside, these fare increases were substantial. The chaos and uncertainties of competition may also have driven some riders away, particularly where the transition was managed badly. On the other hand, some riders almost surely gained, especially those in areas with new minibus services and intense competition.To the extent that riders lost, moreover, the blame seems to lie largely with subsidy cuts rather than deregulation and privatization. Deregulation probably helped the bus industry absorb the government's expenditure cuts, allowing smaller fare increases and ridership losses than would otherwise have been possible, especially in those areas where the competition among bus firms was relatively strong or local authorities and bus companies managed the transition to a deregulated environment relatively smoothly.

Labor probably also lost, although the record is ambiguous. Those employed before privatization seem to have suffered relatively less than might be expected. While the old employees made concessions in work rules and productivity, much of the attrition in the labor force of the large incumbents was accomplished voluntarily, in that workers took early retirement or quit in return for large severance payments. The basic wage rates of the old employees who remained on the job have also been protected in many cases through the negotiation of two-tier wage structures.

Three years after the reforms took effect, total employment in the industry remained about the same, especially among drivers (because bus mileage increased).[34] The newly hired workers are typically paid less than the old workers, as they work for new entrants or as minibus drivers or second-tier big-bus drivers on the old public companies. But if subsidies had been cut without accompanying privatization, the bus companies probably could not have afforded to hire all these new workers, since bus mileage almost surely would have been cut rather than increased. The trade-off for new workers has thus probably been more jobs but at lower wages.

On balance, Britain as a whole has probably benefited from the 1986 bus reforms (assuming no great disparities in the marginal utility of money to those losing and those gaining from the policy). Total bus service is constant to slightly up while costs seem to have declined. The

34. P. Michael Heseltine, "Impact of Deregulation and Privatization of the U.K. Bus Industry on the Labor Force," paper presented at the Conference on Deregulation and Privatization in Passenger Transportation, Tampere, Finland, June 1991, p. 9; and Gómez-Ibáñez and Meyer, *Deregulation and Privatization of Urban Bus Services*, pp. 95–104.

gains to taxpayers probably exceed the losses to labor, since labor's losses seem relatively modest. But whether passengers overall have suffered a loss or gain remains an open question. On the one hand, the fare increases paid by passengers clearly represent a transfer from passengers to taxpayers or bus companies, at least where services have remained the same rather than improved. On the other hand, even the most pessimistic observers acknowledge that service improved in some cases, although for how many passengers and by how much is a hotly debated issue. Indeed, the biggest complication in ascertaining the winners and losers from the British experiment derives from difficulties in valuing the changes in services brought about by the reforms.

Many of the lessons of the British experience are similar to those from the developing countries, particularly the importance of competition. Among the seven metropolitan counties, for example, the reforms seemed more beneficial where competition was stronger, especially for the passengers and taxpayers. The developing countries and British experiences may differ, however, in that competition and subsidies coexist quite happily in Britain. Indeed, the subsidized services seemed to promote competition in Britain by providing a somewhat protected niche for new entrants. Administering the subsidy program has proven challenging for British authorities, however, and whether their experience could be duplicated in other contexts is unclear.

The most distinctive lesson from the British experience, however, is the potential importance of the innovations in service that might be stimulated by privatization. Often the debate over privatization focuses exclusively on the cost savings that might be gained and ignores the possibility of stimulating more market- or customer-oriented services. In Britain, however, service improvements—especially minibuses and new express operations—are in many ways the most dramatic, and to some extent unexpected, consequences of reforms. To the extent, moreover, that the British reforms have created a more entrepreneurial industry and management style, these benefits may persist and grow with the years. While not as dramatically technologically advanced as the innovations unleashed in many parts of the world by telecommunications deregulation, these bus service improvements are nevertheless significant and in the long run could well be one of the strongest arguments for regulatory reform and privatization of transit operations, especially in the developed countries.

Ironically, the potential for privatization-induced bus service innovations is probably more limited in the less developed countries; bus operations in the less developed countries have long displayed much more diversity and entrepreneurial flair than those in the industrialized countries, as explained in chapter 3. Of course, without the spurs of privatization and competition, bus operators in the developing world could fall into the same lethargy as many of their industrialized counterparts. This possibility cannot be ruled out since much of the basic privatization-regulation cycle described in chapter 2 seems impervious to the stage of economic development.

Chapter 5. The U.S. Experience

THE UNITED STATES HAS HAD far less recent experience with privatization and deregulation of urban bus services than Britain or the developing countries. In this chapter we examine this U.S. experience and consider whether the more radical market reforms adopted by Great Britain and some developing countries might work in the United States. We conclude that privatization and deregulation offer many of the same benefits in the United States as they do in other countries, but the problems of implementing these policies in the United States are slightly different than elsewhere and in some respects more difficult.

Transportation and the Private Sector

Although publicly owned and subsidized transit firms are now the norm in the United States, private firms too are active in several ways. For the past several decades, the rarest form of private sector involvement has been the privately owned firm that receives no direct government subsidies. Privately owned unsubsidized firms provided almost all U.S. transit service in the first half of the twentieth century, but most approached or actually entered into bankruptcy and were taken over by public authorities in the 1950s and 1960s. The few private and unsubsidized survivors tend to operate in specialized market niches in a handful of U.S. cities. Private companies operating special premium express bus services survived without subsidies in a few major cities until the late 1970s, for example, although most have been taken over or subsidized since then.[1] Private unsubsidized minibus or van services are active in New York and

1. For a history of private express bus services in New York, for example, see E. S. Savas, Sigurd Grava, and Roy Sparrow, *The Private Sector in Public Transportation in New York City: A Policy Perspective*, report prepared for the U.S. Department of Transportation by the Institute for Transportation Studies, The City University of New York, January 1991, pp. 83–86; or Sigurd Grava, "The Express Bus Saga," *New York Affairs*, vol. 5 (Spring 1979), pp. 111–23.

Miami, often unlicensed and actively suppressed by public authorities.[2] Limousine or bus companies providing specialized services to and from airports are private and unsubsidized in most cities, as are the companies providing most intercity bus services.

A second form of private involvement, also comparatively rare, is the private management company that manages a bus company owned by public authorities.[3] As of the 1980s, such firms managed about eighty bus operations, mainly smaller properties in southern cities.[4] Typically the public authority owns the vehicles and other facilities and employs the drivers and maintenance personnel while the management service company provides a resident team of managers, technical support from the firm's central office, and control over the day-to-day operations of the firm. The management company's discretion and incentive to improve productivity or tailor services to market demands are limited, however, since it typically does not set fares or choose routes or play a key role in labor negotiations.[5]

The most common form of private involvement in urban bus services is contracting with private companies for drivers or special services. While local school authorities have long contracted with private firms for special school bus services in the United States, contracting for transit services for the general public was rare until the late 1980s. The first big wave of contracting out was stimulated in 1973, when Congress passed a law requiring that "no otherwise qualified handicapped individual . . . shall, solely by reason of his handicap, be excluded from participation in, be denied the benefits of, or be subjected to discrimination under any program or activity receiving federal assistance." Public transit authorities attempted to meet this requirement in various ways, often including equipping a portion of their bus fleet with ramps or lifts for wheelchairs or buying buses that "kneeled" at stops to make boarding and alighting

2. See Savas, Grava, and Sparrow, *The Private Sector in Public Transportation*, pp. 86–94; or Allen Randolf, "The Transit Authority's Forced Busing Program," *NY: The City Journal*, vol. 1 (Autumn 1991), pp. 11–13.

3. The majority of these management contracts are held by several companies that in the 1930s and 1940s, when the industry was more profitable, had been holding companies owning, as well as managing, numerous urban public transportation systems.

4. James L. Perry and Timlynn T. Babitsky, "Comparative Performance in Urban Bus Transit: Assessing Privatization Strategies," *Public Administration Review*, vol. 46 (January–February 1986), p. 58.

5. For a description of the management service companies and their contracts see David B. Vellenga, "Can Professional Transit Management Improve Our Urban Transit Systems?" *Proceedings of the Transportation Research Forum*, vol. 14, no. 1 (1973), pp. 345–50.

easier for the elderly and infirm. These specially equipped buses were expensive to buy and maintain and did not solve the problems the elderly and the handicapped had in getting to and from bus stops; most public transit authorities therefore supplemented these measures by instituting special dial-a-ride services for the elderly and the handicapped. Dial-a-ride typically uses small lift-equipped vans and offers door-to-door service with a day or two advance reservation required.[6] Many public transit authorities contracted with private vendors for these services because they were labor intensive and therefore costly to provide with regular public bus drivers.[7]

A second wave of contracting out began shortly after President Ronald Reagan took office in 1981, when the Federal Transit Administration (FTA), the agency of the Department of Transportation that administers federal transit aid, began promoting privatization as a means to control the rapid inflation in transit costs that had occurred in the 1970s.[8] Between 1970 and 1980, for example, real operating expenses for the U.S. urban mass transit industry grew from $3.6 billion to $6.0 billion (measured in constant 1980 dollars) while farebox revenue declined from $3.2 billion to $2.6 billion and the deficit covered by public subsidy increased from $0.3 billion to $3.5 billion. While fares failed to keep pace with inflation and ridership continued to drop even as service expanded, 43 percent of the increase in the deficit occurred because of rising real labor costs per vehicle-mile of service operated.[9]

In 1984 the FTA required that public transit authorities receiving federal aid demonstrate that they considered "the private sector's capacity to provide needed transportation services" not only through contract-

6. For an overview of the evolution of these policies and dial-a-ride see John R. Meyer and José A. Gómez-Ibáñez, *Autos, Transit and Cities* (Harvard University Press, 1981), pp. 230–34, 73–76.

7. For a description of different contracting arrangements used, see Sandra Rosenbloom, "Role of the Private Sector in the Delivery of Transportation Services to the Elderly and the Handicapped in the United States," *Transportation Research Record*, no. 1170 (1988), pp. 39–45.

8. The Federal Transit Administration was called the Urban Mass Transportation Administration (UMTA) until 1991; for the sake of simplicity its present name is used here.

9. The balance was because of increases in energy prices (10 percent), increases in the number of vehicle-miles of service operated (16 percent), declines in the number of passengers carried per vehicle-mile of service (2 percent), and reductions in real fares (28 percent); see Don H. Pickrell, "Rising Deficits and the Uses of Transit Subsidies in the United States," *Journal of Transport Economics and Policy*, vol. 19 (September 1985), pp. 281–98.

ing out for subsidized services but through allowing unsubsidized private services to operate as well.[10] At about the same time, the FTA announced a program of demonstration projects in which local authorities, with the support of special federal grants, could experiment with and evaluate various forms of private sector involvement. By 1988 the demonstration program included seven projects, four of which included contracting out for regular route bus services in Miami (forty buses), New Orleans (twenty-three buses), Los Angeles (seventy buses), and Cincinnati (seventeen buses). The remaining three projects, in Des Moines, Minneapolis, and Little Rock, were more ambitious, inspired in part by the British example. In these three projects, a separate public authority was established, contracting service from private or public vendors on a competitive basis.[11]

Many local officials and their representatives in Congress viewed the Reagan administration's privatization initiative with skepticism. Some saw it as a usurpation of traditional local prerogatives, while others regarded it as an attack on transit labor or a thinly disguised rationale for reducing federal transit aid, which the Reagan administration was also advocating at the time. In 1987 a conference committee of the two houses of Congress, in recommending a transit appropriations bill that greatly exceeded the administration's request, rejected the possibility, which the FTA was considering at the time, that local authorities be required to contract out a minimum percentage of their services as a condition for federal aid.[12]

Most contracting out seems to have been initiated to serve local needs rather than to meet federal requirements. Cost savings seem to have been the primary motivation. Contracting out is prevalent among smaller communities in California, for example, in part because the state's transit aid program is designed to give local authorities a great deal of incentive and opportunity to be cost conscious; a share of the state's transit aid is

10. "Private Enterprise Participation in the Urban Mass Transportation Program," *Federal Register*, vol. 49 (October 1984), p. 4310.

11. At the same time UFTA also began demonstrations of contracting out for the maintenance of vehicles; see Department of Transportation, Urban Mass Transportation Administration, "Summary Descriptions of Fully Competitive Transit, Competitive Services, Maintenance Contracting, and Financial Infrastructure," Washington, 1987.

12. Congress, *Making Appropriations for the Department of Transportation and Related Agencies*, conference report to accompany H.R. 5205, 100 Cong. 2 sess. (Government Printing Office, 1987), pp. 28–29.

delivered directly to cities and towns rather than to regional public transit authorities and can be used for road improvements as well as bus service.[13] In other cases, capacity to initiate service or an ideological commitment to private enterprise seems to have been the primary motive for contracting out. Houston's regional transit authority contracted out most of its express bus service after voters increased transit funding and the authority thought contracting was the fastest means to meet the public's desire for expanded service.[14] In 1988 a conservative Colorado state legislator successfully sponsored a bill requiring Denver's regional public transit authority to competitively contract 20 percent of its service.

Nevertheless, a 1985 survey revealed that contracting out accounted for a relatively small proportion of total operations and was usually confined to specialized and supplementary services. Contracted services accounted for only 5.1 percent of the total nationwide transit operating expenditures on fixed-route or dial-a-ride services and only 8.6 percent of the total revenue vehicle-miles of services offered. Dial-a-ride services alone constituted 58 percent of the contracted services in the survey.[15]

The Results of Contracting Out

Despite its limited application, contracting out is the best source of information about the potential cost savings from privatization in the United States. During the past several decades many researchers have used statistical comparisons of the costs of U.S. private and public bus firms to attempt to determine whether private bus firms are less expensive or more efficient than their public counterparts. Most of these studies found private firms were more efficient than public, but others found the opposite or no difference.[16] All of these studies have found it difficult to control for the other factors, besides ownership, that might influence costs, such as differences in services provided or the local operating

13. Roger F. Teal, "Transit Service Contracting: Experiences and Issues," paper presented at the sixty-fourth annual meeting of the Transportation Research Board, Washington, January 1985, pp. 2–3.

14. Genevieve Guiliano and Roger F. Teal, "Privately Provided Commuter Bus Services: Experiences, Problems, and Prospects," in *Urban Transit: The Private Challenge to Public Transportation*, ed. Charles A. Lave (San Francisco: Pacific Institute for Public Policy Research, 1985), p. 162.

15. Roger F. Teal, "Public Transit Service Contracting: A Status Report," *Transportation Quarterly*, vol. 42 (April 1988), pp. 211, 212.

16. For one of the latest studies and a brief review of earlier ones see Perry and Babitsky, "Comparative Performance in Urban Bus Transit," pp. 57–65.

environments (for example, traffic congestion or regional wage rates); these comparison problems have become more acute as the remaining U.S. private bus firms have retreated into specialized market niches. In some cases the firms classified as private are also only nominally so; in New York City and Westchester County, New York, for example, several private bus operators offer subsidized service under contract with public authorities, but their contracts are not competitively bid and public oversight and involvement is so great that it would be hard to classify the firms as private in the traditional or strict sense.[17]

Fixed-Route Systems

Contracting for fixed-route service through competitive bidding is still a fairly recent phenomenon in the United States, and the practice is still evolving and rather poorly documented.[18] The typical contract affects only a small part of the public agencies' fixed-route service, usually less than 10 percent. Contracts for peak hour or commuter express bus services are slightly more common than for all-day services. The public agency often supplies the vehicles, since federal grants of up to 80 percent are available for vehicles purchased by public authorities. The private contractor supplies drivers and, most of the time, is responsible for supervision and routine vehicle maintenance as well. Contracts of one to three years in length are typical, and sometimes the contracts are apparently renegotiated rather than put out for competitive bids when they are up for renewal. Little information is available on the numbers of bids received or the types of private firms that bid for the contracts, although many contractors seem to be drawn from either private scheduled or intercity charter bus operators or private companies that already operate special school bus services under contract to local governments.[19]

17. For a description of the New York and Westchester arrangements see URS Company (in association with Polytechnic Institute of New York), *Express Bus Route Policy Study*, report prepared for New York City Department of City Planning (October 1986).

18. The best sources of information about contracting practice are a series of brief fact sheets produced by the Rice Center of Houston, Texas (under the title *Public Sector Briefs)* and a series of articles and reports to the FTA by Roger F. Teal. See especially "Transit Service Contracting," "Public Transit Service Contracting," *Transportation Quarterly*, vol. 42 (April 1988), pp. 207–22, and "Issues Raised by Competitive Contracting of Bus Transit Service in the USA," *Transportation Planning and Technology*, vol. 15, no. 2 (1991), pp. 391–403.

19. For a survey of typical contract size and terms as of 1985 see Teal, "Public Transit Service Contracting."

TABLE 5-1. Cost Savings in the United States When Private Contractors Substitute for Public Agencies on Fixed Routes

Public agency sponsor	*Number of buses*	*Cost savings over public service (%)*
Bay Area Rapid Transit District	45	26
City of Los Angeles, California	19	32
Fairfax County, Virginia	33	39
Fort Wayne, Indiana	86	22
Johnson County, Kansas	21	39
Snohomish County, Washington	58	22
Yolo County, California	14	37

SOURCES: Data supplied by public agency sponsors as cited by Roger F. Teal, "Transit Policy Viewed from a Fairy Tale World: A Critique of *The Emperor's New Clothes*," University of California, Irvine, October 1989, p. 5; the data originally appeared in Rice Center, *Private Sector Briefs* (Department of Transportation, Urban Mass Transportation Administration, 1985–88).

The available evidence suggests that contracted services cost 20 percent to 30 percent less than comparable services provided by the local public bus company, savings similar to those reported from the British experience with contracting out in London. A study of seven fixed-route services, summarized in table 5-1, for example, found cost savings between 22 and 39 percent, with an average savings of 31 percent. The costs to public agencies of administering contracts were not considered in the estimates; these administrative expenses vary considerably, but average around 6 percent of contract costs,[20] which would mean the savings net of administrative costs would average around 25 percent in the seven services examined. A detailed study of the contracting experience in Denver found that private operator costs, including the cost of public agency contract administration, were 17 percent less than the "incremental" operating costs of the public bus company; private costs were 27.5 percent less than the "fully allocated" costs of the public bus company, which include a portion of its overhead and depreciation.[21]

Only a modest portion of any savings seems to come from lower wages for drivers, with the rest from lower fringe benefits, higher driver productivity, and lower overhead and supervision costs. A study in Dallas

20. Teal, "Issues Raised by Competitive Contracting," pp. 396–97.

21. These figures are estimates for a "stable year" based on the first year of contracting experience. In the first year the savings were less pronounced because the Colorado legislature had required Denver's public bus company to contract 20 percent of its bus service without laying off public employees, so the public company was paying for some unutilized employees until natural turnover and attrition reduced its work force; see KPMG Peat Marwick, *Performance Audit of Privatization of RTD Services*, revised final report prepared for the Regional Transportation District (Denver, Colorado, December 24, 1990).

showed, for example, that 17 percent of private sector savings occurred because of lower wages for drivers, 45 percent from lower driver fringes, 27 percent from lower overhead, and 11 percent from lower maintenance costs.[22] Fringe benefits may be an important source of savings because U.S. public sector bus driver fringe benefits average 50 percent of wages, double that of the average private sector worker.[23]

Contracting out also seems to induce reductions in the costs of some local public bus company operators, much as it did in London. In Portland, Oregon, for example, public employee unions granted wage and work rule concessions in 1985 in return for a management promise not to expand contracting of fixed-route services for a three-year period, while in San Diego, California, and the Tidewater area of southeastern Virginia the public agencies gained similar concessions without any loss in contracting freedom.[24]

Some public agencies and labor unions have questioned these estimates of cost savings.[25] One concern is that the private contractors' initial bids may be unrealistically low, because the contractors lack experience or because they hope that, once they have won the initial contract, future contracts will be negotiated at higher prices and perhaps without competitive bidding. So far, the U.S. experience, like that of Britain, indicates that the savings generally do not disappear over time. One analyst examined twelve U.S. services contracted for four years or more by ten different public agencies and found that real unit costs had declined over time on nine services and increased in only three. In those

22. The total savings for private versus public costs in Dallas was 30 percent; see Teal, "Issues Raised by Competitive Contracting," p. 402.

23. Jean Love and Jim Seal, "Competitive Contracting in the U.S.: Overcoming Barriers," paper presented at the Second International Conference on Privatization and Deregulation in Passenger Transportation, June 1991, p. 5.

24. Rice Center, *Private Sector Briefs*, vol. 1, 1986–1988 (Houston), pp. 13-1, 13-2, and 13-5.

25. These questions have been raised most forcefully by several academics and consultants in Elliott D. Sclar, K. H. Schaeffer, and Robert Brandwein, *The Emperor's New Clothes: Transit Privatization and Public Policy* (Washington: Economic Policy Institute, 1989); and Elliot D. Sclar, "Less Than Meets the Eye: Colorado's Costly Experience with Transit Privatization," Washington, Economic Policy Institute, January 1991. For replies by proponents of privatization see Roger F. Teal, "Transit Policy Viewed from a Fairy Tale World: A Critique of *The Emperor's New Clothes*," University of California, Irvine, October 1989; and Wendell Cox and John Love, "A Public Purpose for Public Transit: A Response to the EPI Report," Policy Study 207 (Santa Monica: Reason Foundation, January 1990).

three cases, costs rose by less than the average rate of public transit cost inflation.[26]

Another objection is that private contractors will offer less reliable and lower-quality service. Private bus drivers are likely to be less skilled and dependable than public drivers because they are paid less, some have argued, and for-profit contractors will attempt to reduce costs by cutting quality unless they are closely monitored.[27] These criticisms gained credibility because of widely publicized problems in two of the FTA's fixed-route demonstration projects: New Orleans dismissed its initial contractor because of unreliable service, and Miami has threatened to cancel unless service is improved. In Miami and New Orleans the contractor was Greyhound, the dominant intercity bus operator in the United States. Greyhound acknowledged that its inexperience in urban bus services contributed to the problems, but it also blamed the New Orleans and Miami public agencies for supplying it with the least reliable buses in the agencies' fleets.[28]

Finally, critics have argued that the savings are calculated in a manner that exaggerates the costs of the public sector alternative and understates the costs of the private contractor. A particular focus has been the FTA's requirement that agencies receiving federal aid compare private contractor bids with the "fully allocated" costs of publicly operated service. Critics contend that fully allocated costs unfairly burden the public operators with overhead functions, such as general administration or route planning, that would not be avoided by the use of private contracting. The FTA imposed this requirement because of concern that public bus companies might understate their costs by allocating some of the costs associated with services that were being competitively procured to services that were not.[29]

26. In two of the three cases, moreover, the contracts for the services were renegotiated rather than rebid during the period covered; see Teal, "Issues Raised by Competitive Contracting," p. 395.

27. See, for example, Sclar and others, *The Emperor's New Clothes*, p. 29.

28. Telephone interview with Douglas Birnie, chief, Office of Private Sector Initiatives, FTA, April 1990; and Teal, "Issues Raised by Competitive Contracting," pp. 399–400.

29. Despite the critics' objections, the official guidelines for full cost allocation, which were drawn up by a competitive services advisory board composed of accounting firms and public and private bus companies, seem to strike a sensible balance. The guidelines say to exclude from the estimates of fully allocated costs certain overhead functions that remain with the public company, such as service planning, and to add to the private sector bids the costs of new functions that the public company might have to assume, such as contract monitoring. Price Waterhouse, Office of Government Services, "Fully Allocated Cost Analysis Guidelines for Public Transit Providers," report prepared for the Urban Mass Transportation Administration (Department of Transportation, April 1987).

Dial-a-Ride Systems

A serious problem in comparing public and private dial-a-ride costs is that literally hundreds of very small firms are involved, each one using different cost-accounting conventions and providing different types and qualities of service. Some dial-a-ride services carry ambulatory riders only, while others carry nonambulatory riders (for example, wheelchair users). Some services provide rides from the patron's home to selected destinations only, such as local hospitals, shopping malls, social service agencies, or churches, while others will serve any destination the rider selects within a general service area. While some systems require at least one- to two-day advance reservations, others will consider last-minute requests for service on a space-available basis.

One study of seventy dial-a-ride services suggests that private costs are not always lower than public costs. Sandra Rosenbloom adjusted the costs of the systems to include any overlooked items and grouped the services into five types. As shown in table 5-2, Rosenbloom found that the ranges of private and public sector costs overlap considerably for each of the four types of service that both sectors provide, although in three out of four cases both the least expensive and the most expensive public systems are significantly more costly than their private sector least and most expensive counterparts. Furthermore, the fifth and strictly private system, the taxicab, seems by many measures to be the cheapest of all methods of servicing these needs, although taxicab service may also be inferior for many users.

The study argues that the potential cost advantages of the private sector are often not realized because the local public authorities contract in a way that reduces competition in bidding. Often these discouragements are unintentional, but sometimes local authorities appear to deliberately manipulate the process to limit competition, perhaps with the idea that a single large contractor will be easier for the authority to monitor.[30]

Detailed interference in day-to-day operations may also limit the advantages of private sector contracting. Local authorities often maintain separate centralized scheduling and dispatching systems, for example, even when a single contractor is selected to provide all the dial-a-ride service. Local authorities find centralized scheduling and dispatch appealing because it provides the opportunity to group trips and, if publicly

30. Sandra Rosenbloom, "Role of the Private Sector in the Delivery of Transportation Services to the Elderly and the Handicapped in the United States," *Transportation Research Record*, no. 1170 (1987), p. 40.

TABLE 5-2. Costs per Trip on Publicly Operated and Privately Contracted Dial-a-Ride Services for the Elderly and Handicapped
1986 dollars

Type of service	*Publicly operated*	*Contracted to private firm*
Ambulatory patrons, to congregate destinations	3.80–6.90	4.20–11.00
Ambulatory patrons, to independent destinations	12.00–18.00	6.30–11.00
Nonambulatory patrons, to congregate destinations	14.50–29.00	9.90–17.90
Nonambulatory patrons, to independent destinations	14.00–31.40	11.10–27.10
Taxi voucher systems	n.a.	5.10–8.40

SOURCE: From an analysis of seventy systems by Sandra Rosenbloom, "Role of the Private Sector in the Delivery of Transportation Services to the Elderly and Handicapped in the United States," *Transportation Research Record*, no. 1170 (1987), p. 43.2.
n.a. Not available.

operated, is a means of monitoring and controlling the activities of the private contractor. There are fewer opportunities for group trips than are popularly imagined, however, and any savings from greater vehicle and driver productivity is usually more than offset by the costs of the central dispatching operation. Some evidence also exists that centralized dispatching actually reduces productivity because the private operators are experienced at scheduling and more in touch with their fleet and drivers than the centralized dispatch office.[31]

Lessons and Contrasts: Britain versus the United States

The gradual and piecemeal privatization through contracting out in the United States contrasts sharply with the sudden and wholesale privatization that occurred in Britain in 1986 and in some of the developing countries. Not only has the pace been slower in the United States but only a small fraction of services are contracted in most instances. Furthermore, there has been a reluctance in the United States to experiment with more complete forms of privatization, particularly "commercialized" (that is, unsubsidized and deregulated) private bus services. The slower, less comprehensive, and more cautious U.S. approach has pros and cons, as a contrast of the U.S. and British experiences reveals.

31. Rosenbloom, "Role of the Private Sector," p. 42.

The Pace of Reform: Gradual versus Immediate

A sudden and complete shift to contracted or privatized services avoids the difficult problems of making certain that public bus companies are not competing unfairly with private bus companies for contracted service. Complete privatization eliminates the need to monitor public sector bids, for example, by eliminating any subsidy that is not competitively procured.

A more sudden and complete privatization may also evoke greater cost savings, by making it much easier to shed excessive overhead, to renegotiate labor agreements, and to generally rethink operations in fundamental terms. Some weak but suggestive evidence that this might be true is that the proportionate cost savings of new British entrants (outside of London), or of privatized firms in developing countries, is often greater than that from partial contracting out reported in London or the United States.

A gradual approach, however, avoids the transitional problems experienced in Britain outside of London in 1986. Even with a year's notice and preparation, the overnight transition from a largely public sector to a privatized one caused confusion, even chaos, that almost surely drove ridership down, some fear permanently.

The gradual approach also allows public authorities to experiment with contracting procedures to avoid some of the contracting problems experienced in the first months after the reforms in Britain or with the dial-a-ride programs in the United States. With a more gradual approach, public authorities can, at least in theory, develop contracting procedures that encourage new entrants and competition. With no requirement to completely privatize, public authorities could also employ public companies in those specialized market niches where competition is too weak or the problems of contract monitoring and enforcement too difficult.

Finally, the experience with gradual and partial contracting in the United States and London suggests that the presence of competition is more important than wholesale privatization in evoking cost savings. If the public bus companies perceive themselves as competing on fair terms with private bus companies, for example, then across-the-board privatization may not be necessary to achieve across-the-board cost savings. London reports that contracting out less than 20 percent of services stimulated cost reductions in the regular public bus company, and several U.S. cities, such as San Diego and Portland, report similar experiences. The key seems to be not the amount of contracting but whether the employees and managers of the public bus company believe that local authorities are willing to contract more services if the cost advantage of the private over the public sector is substantial.

Contracting Out versus Commercialization

The principal disadvantage of relying exclusively on contracting out is that the services provided are designed by the public rather than the private sector. Thus while contracting out may evoke cost and productivity savings, it is less likely to encourage the service innovations, such as route changes or high-frequency minibus services, that have been such striking and important features of privatized bus firms in Britain and many of the developing countries. Of course, the innovational advantage of commercialization would be of little value in the United States if few or no U.S. transit services were commercially viable. What, then, are the prospects for successful commercializations and related innovations in the United States? Obviously Britain's experience is likely to be more relevant than that of the less developed countries because of the greater similarities in the British and American situations.

Prospects for commercial viability. One of the most striking results of British bus deregulation is that more than 80 percent of the bus mileage operated before deregulation was commercially registered for operation without subsidies after deregulation; this figure even surprised most deregulation proponents.[32] The commercially viable part of the route network would almost surely be smaller in the United States than in Britain and would likely focus on different types of services.

The problem is not that the cost savings or productivity improvements from privatization might be smaller in the United States than Britain. Indeed, to the extent that contracting out is a guide, the unit cost reductions possible in the United States and Britain seem remarkably similar (about 20 to 30 percent). Comparisons of wage rates and productivity in British and U.S. public bus firms, though somewhat difficult to make, support this result.[33] The opportunities for wage and benefit con-

32. In Britain, commercially registered services still receive some subsidies in the form of reimbursements for concessionary fares, but these reimbursements should not affect profitability greatly since in most instances they amount to only about 15 percent of total bus costs. R. J. Balcombe, and others, *Bus Deregulation in Britain: A Review of the Opening Stages*, Report 107 (Crowthorne, England: Transport and Road Research Laboratory, 1987), pp. 1, 3.

33. These comparisons are explained in detail in José A. Gómez-Ibáñez and John R. Meyer, *Deregulating and Privatizing Urban Bus Services: Lessons from Britain*, report prepared for the Urban Mass Transportation Administration (Department of Transportation, January 1989), pp. 163–78.

cessions seem comparable because wages for bus drivers in Britain before deregulation were similar to those in the United States as measured by the relation between wages for bus drivers and the average wages for other types of labor.[34] The productivity of U.S. bus companies is also roughly similar to that of their British counterparts before deregulation, although the United States may enjoy a slight advantage when productivity is measured in the number of bus miles or hours of service operated per unit of input (for example, labor or capital) while the British may enjoy an advantage when productivity is measured by the number of passengers carried per unit of input.[35] This would imply that the opportunities for productivity improvement in the United States lie slightly less in improving the efficiency with which bus miles of service are produced than in pruning or redeploying the mileage to more closely match passenger demands. Since U.S. transit demand is more highly concentrated in the peak hours, moreover, there may be greater scope for improving labor productivity in the United States than in Britain through the relaxation of work rules, such as limitations on part-time and split-shift labor.

The key obstacle to achieving British levels of commercialization in the United States is that British bus service has always been less heavily subsidized than U.S. service, and therefore always closer to commercial viability. Because of the lack of good depreciation or capital expense data, only the ratio of farebox revenues to operating expenses *excluding* depreciation can be calculated and compared for both countries.[36] This ratio averaged between 50 percent and 75 percent for the British metropolitan counties outside London before deregulation (with the exact percentage depending considerably on whether concessionary fares to

34. Comparisons of fringe benefits were not possible; the omission of fringes may understate the savings in the United States since U.S. public bus driver fringes seem very high.

35. These comparisons are based on prereform data from the public bus companies of seven British metropolitan areas (excluding London) and data from sixteen major bus firms serving comparable U.S. metropolitan areas. Productivity was measured in various ways, including adjustments for differences between the United States and Britain in the average operating speeds, passenger trip lengths, vehicle seating capacities, and average hours worked each week.

36. Perhaps because capital expenditures are almost wholly financed out of federal government grants in the United States, most U.S. transit firms, unlike their British peers, do not calculate depreciation or include it or other capital charges in their reported operating expenses. As in Britain, U.S. transit firms are also exempt from fuel taxes, although these taxes averaged only 19.6 cents a gallon in 1984, or about one-fifth the British rates; Bureau of the Census, *Statistical Abstract of the United States, 1987* (Washington: Government Printing Office, 1986), p. 583.

school children, the elderly, and so on were counted as fares or subsidies). Comparable farebox recovery ratios for U.S. transit firms average a little less than 40 percent.[37]

Britain's lower subsidies before the reforms were not the result of higher fares. Bus fares in the British metropolitan counties averaged seventeen pence around 1985 whereas the average bus fare for the thirty-one largest all-bus U.S. transit firms averaged thirty-three cents in 1983, or approximately twenty-two pence in 1985 pounds.[38] British fares were somewhat more akin to U.S. fares in relation to average family purchasing power, since Britain's gross national product per capita in 1985 was only U.S. $8,000 compared with the U.S. figure of U.S. $16,700.[39] Nevertheless, the prospects for commercialization were almost certainly better in Britain before the 1986 reforms than in the United States: in Britain, fares were lower and additional revenue needs were less pronounced.

Britain's lower fares and subsidies probably occurred largely because of its significantly higher level of transit ridership and more favorable pattern of transit use. The British averaged 114 transit rides per capita (100 by bus) nationwide in 1985–86, or about three times the U.S. average of 34 rides per capita (23 by bus) in 1983. Transit ridership per capita is much higher than the national average in the metropolitan areas of both countries, but the relative ridership rates remain about the same. Most of the major British metropolitan areas average around 200 rides per person per year; the largest and oldest U.S. metropolitan areas (such as Chicago, San Francisco, or Boston) average around 100 rides per person per year, with most of the remaining U.S. metropolitan areas averaging about 50 or fewer.[40]

Britain's higher levels of transit ridership reflect much lower levels of automobile ownership. In 1984 there were 305 cars and taxis and 365 motor vehicles of all types per thousand population in the United Kingdom, versus 540 cars and taxis and 722 vehicles of all types in the United States.[41] Those British households that do own a car or truck are much more likely to own only one vehicle rather than two or more. In the late

37. Gómez-Ibáñez and Meyer, *Deregulating and Privatizing Urban Bus Services*, pp. 143–47.

38. Gómez-Ibáñez and Meyer, pp. 147–50.

39. Bureau of the Census, *Statistical Abstract*, p. 825.

40. Gómez-Ibáñez and Meyer, *Deregulating and Privatizing Urban Bus Services*, pp. 150–53.

41. U.K. Department of Transport, *Transport Statistics of Great Britain, 1975–1985* (London: Her Majesty's Stationery Office, 1986), p. 182.

1970s in Britain 33 percent of all households had no car or truck and 13 percent had two or more, while in the United States those figures were more than reversed with 15 percent having no car or truck and 50 percent having two or more.[42] Driver licensing is also much lower in Britain, particularly among women, probably because of the relative rarity of households with two or more cars. Only 68 percent of all men and 30 percent of all women in Britain seventeen years of age or older were licensed to drive in 1979 while 90 percent of all men and 78 percent of all women age sixteen or older were licensed in the United States in 1983.[43]

Britain's higher transit ridership may allow bus operators to exploit more economies from traffic density and limit the number of marginally profitable low-density routes. British bus firms may gain an even more important advantage because their denser passenger volumes are also more evenly distributed over the hours of the day and the days of the week, so that British bus operators are less exposed to severe peaking of demand and attendant additional costs than their U.S. counterparts.[44] Britain's less peaked pattern of transit use is probably caused by its auto ownership pattern. The British household's one car is often used for commuting so that the spouse and children must use public transportation for many nonwork trips. Since far fewer British women are licensed to drive, much of the family's shopping and personal business must be done by bus in Britain. Furthermore, as already noted, more than twice as many British households, as a percent of total households, are without cars than in the United States.

The differences between the British and American transit markets suggest where commercialization might be successful in the United States. The U.S. transit market can be segmented into three major components: radial commuting, inner city, and intersuburban. The radial commuting market is dominated by work trips from suburbs to the center of the

42. This disparity is also not explained by differences in average household size: in households with two adults, for example, the difference between U.S. and British car ownership is even greater than among the population as a whole; Gómez-Ibáñez and Meyer, *Deregulating and Privatizing Urban Buses*, p. 154.

43. U.K. Department of Transport, *National Travel Survey: 1978/9 Report* (London: Her Majesty's Stationery Office, 1983), p. 4; Federal Highway Administration, *Personal Travel in the United States: 1983–1984 Nationwide Personal Transportation Study*, vol. 1 (Department of Transportation, 1986), p. 4-3.

44. While bus and rail transit capture four times the share of commuting trips in Britain as in the United States (21 percent vs. 4.5 percent), they capture 15 times the shopping and personal business trips (15 percent vs. 1.1 percent) and six times the share of social trips (12 percent vs. 1.8 percent); Gómez-Ibáñez and Meyer, *Deregulating and Privatizing Urban Bus Services*, pp. 155–58.

metropolitan area, although in a few metropolitan areas trips to other large and congested subcenters may also be important. The inner-city market is service to and within the higher-density residential neighborhoods surrounding the metropolitan center, including communities that may be relatively well-to-do as well as low-income white and minority neighborhoods. The intersuburban market serves trips within and between the lower-density and more auto-oriented suburbs on the edge of the metropolitan area.

Each market has a unique demand pattern that influences its potential profitability and the operating strategies that a bus company must adopt to serve it successfully. The inner-city market is most like the typical British bus market: services to dense, often poorer neighborhoods with a lower level of automobile ownership, leading to higher volumes of bus passengers in a pattern that is relatively heavy and more evenly distributed over the day. By contrast, the radial commuting market, while often dense, is also highly peaked. The intersuburban markets include some commuting trips, but most patronage comes from the small minority of suburban residents who do not have access to an automobile, including those too young or old to drive; as a result, intersuburban markets have a more moderate level of peaking than radial commuting markets but an extremely low density of travel demand.

The inner-city market seems to account for as much as half of all patronage on many U.S. bus transit systems, with radial markets a close second and intersuburban service a distant third. Data on farebox recovery ratios by route or market are not generally available, but the few that are support the strong consensus in the U.S. transit industry that inner-city routes have the highest farebox recovery ratios with radial commuting routes second and intersuburban routes lagging badly. Express radial commuter bus services are something of an exception to this rule since they often charge a premium fare and are sometimes as profitable as inner-city services.[45]

45. Data from an anonymous large metropolitan transit authority in the early 1970s, for example, show that fare revenues were 133 percent of operating costs on its inner-city and short radial routes, 102 percent on longer radial, and only 66 percent to 75 percent on suburban and crosstown services; Meyer and Gómez-Ibáñez, *Autos, Transit and Cities*, p. 53. Similarly, the San Antonio transit authority, which has been following a strategy of heavily subsidizing its services, covered 29 percent of its costs out of the farebox on inner-city and radial routes in 1986, but only 10 percent on suburban services in that same year; Christian Young, planner, VIA Metropolitan Transit Authority, San Antonio, personal communication, August 14, 1987.

Clearly, the potential for commercialization in the United States is strongest in the inner-city markets. There is probably also a strong opportunity for commercializing many of the radial commuting services, especially express services that are the major remaining outpost of the private operator in the U.S. transit industry. The market with the least potential for commercialization in the United States is clearly inter-suburban service.

Prospects for service innovation. Privatization and commercialization in the United States might induce service innovations, as in Britain, that expanded the amount or types of services that are commercially viable. Indeed, the historical record in the United States is somewhat encouraging. For example, premium express commuter bus services were pioneered in New York in the late 1960s by that metropolitan area's few remaining private companies; New York's public transit agency copied the private innovation only after several years of demonstrated success.[46]

Moreover, there seem to be some underexploited opportunities to improve U.S. bus profitability by tailoring services and fares to market demands. For example, one of the most striking differences between British and U.S. bus operations, even before the advent of British bus deregulation, was the greater complexity and differentiation of British bus fares relative to American. British fares vary much more by distance traveled: fares typically increase in about six to eight stages from fifteen pence to twenty pence for a trip of 0.5 mile to fifty pence to eighty pence for a trip of around 10 miles.[47] In the United States barely more than half of the large metropolitan areas have bus fares that graduate with distance, and in most of those cases surcharges are applied to only a few very long trips or express bus services.[48] Transfers between bus routes are seldom free in Britain but are often free in the United States, which further reduces how much fares vary by distance traveled in the United

46. One of the reasons for the delay was that the public agency, the New York City Transit Authority, was convinced, probably rightly, that the express buses were drawing a large share of their ridership from the authority's own local bus and subway lines. These public services were probably losing money on incremental riders, however, especially in the peak when extra capacity is costly to provide. See Savas, Grava, and Sparrow, *The Private Sector in Public Transportation*, pp. 83–86

47. Kenneth Perrett and J.C. Vickers, "Fares and Ticketing in PTE Areas," Department of Transport, Transport and Road Research Laboratory, Crowthorne, England, Working Paper WP(TP)52, 1987.

48. Don Pickrell, "Variation in Urban Transit Costs and Revenues by Type of Service," paper presented at the sixty-seventh annual meeting of the Transportation Research Board, Washington, January 1987, p. 12.

States. British bus operators are also more likely than U.S. operators to offer some form of discount for off-peak travel.[49]

Obviously, distance-graduated and peak fares should permit operators to tailor their fares to more closely match costs. To the degree that longer-distance and peak-period bus travelers are willing to pay more (because they are wealthier or have fewer attractive alternatives), these fare structures may help bus companies to capture more of the value that passengers place on these services. By contrast, the common U.S. practice of charging a single flat fare for most types of trips may discourage short-distance inner-city travelers who are relatively profitable to serve and encourage long-distance travelers who are currently very unprofitable but who might be willing to pay more of their costs.

Minibuses are another innovation that might find greater use in U.S. transit. Transit operators in Britain found minibuses effective in serving certain markets, especially routes with relatively low volumes of passengers or fairly infrequent service, areas that did not have highly peaked demand periods, and previously untapped residential areas. Cost simulations suggest that much the same potential exists in the United States.[50] Minibuses could be very competitive in the United States for low-volume routes, especially if some labor concessions were made on the use of part-time drivers. For high and medium volumes, however, the conventional fifty-passenger bus appears to be generally better under U.S. conditions.

By introducing minibuses on low-density routes, U.S. transit operators could effectively downscale the size of the vehicle (supply) to match the size of the market (demand). In addition, service frequencies on some routes should increase with minibuses. To the extent that the increased frequency was perceived as a significant improvement in service, ridership might increase, thus making the minibus service even more viable.

However, the likelihood that U.S. transit markets are in general more peaked than Britain's is not advantageous for minibus operations. Minibuses are usually best at serving less peaked markets, where the

49. Usually the British bus operator will have a lower maximum fare in the off peak. In the United States, by contrast, off-peak discounts are so rare that Cevero could only find about two dozen cases to study in the entire country. Pickrell states that in 1981 operators offered an off-peak discount in only one out of twenty-six major U.S. cities. See Robert Cervero, "Examining Recent Transit Fare Innovations in the United States," *Transportation Policy Decision-Making*, vol. 3 (1985), p. 29; and Pickrell, "Variation in Urban Transit," p. 12.

50. See Gómez-Ibáñez and Meyer, *Deregulating and Privatizing Urban Bus Services*, pp. 179–87.

penalty for their relatively high labor costs per passenger carried is reduced. The U.S. inner-city markets seem to have a peak-to-base ratio similar to that observed in Britain, around 1.5 or 2 to 1. Therefore, minibuses might be usefully deployed on some lower-density inner-city or central business district routes in U.S. cities. Suburban routes in the United States probably have a peak-to-base ratio of around 3 or 4 to 1, however, making minibus deployment there less attractive, even though the low densities in these markets otherwise seem to favor minibus operations.

These possibilities are broadly supported by the appearance in New York and Miami during the 1980s of private operators, many of them unlicensed and illegal, providing service with ten- to twelve-passenger vans.[51] Some of New York's private vans serve major subway stations or local shopping and commercial centers in moderate-income, minority neighborhoods in New York's outer boroughs (particularly Queens). They were started by private individuals with automobiles in the 1970s, but as ridership grew, especially after a transit strike in the early 1980s, the automobiles were replaced by passenger vans. Many of the drivers are recent immigrants from the Caribbean, and in some cases they have organized route associations to provide a stronger identity for their route and certain common services, such as off-street loading and unloading areas. Other New York private vans provide long-distance express commuter services to Manhattan, mainly from Staten Island and New Jersey. Some of these vans, especially those that provide interstate service (from New Jersey or from Staten Island via New Jersey) are legal since the federal licenses required for interstate service are easier to obtain than the New York State or city licenses required for noninterstate service.

New York's inner-city van services are strikingly similar to some of the British minibus services. They charge the same fare as the competing local public bus lines but are still attractive to riders because they arrive at stops more frequently and are faster (primarily because they need to make fewer stops to get a full load). The development of commuter vans is somewhat surprising, however, since demand is heavily peaked and the distances long enough that most drivers make only one, or at most two, trips each morning and each night. Despite these disadvantages, the

51. This description of New York's private van services is based primarily on Savas, Grava, and Sparrow, *The Private Sector in Public Transportation*, pp. 86–94; Randolf, "The Transit Authority's Forced Busing Program"; and Metropolitan Transit Authority, Policy and Planning Office, "Van and Car Service Operations Affecting NYCTA Surface Operations," New York, January 1992.

private commuter vans are able to make a profit charging a fare slightly lower than competing public express buses ($3.75 versus $4.00). Moreover, the competing public bus lines are unprofitable.

The potential to introduce minibuses in the United States must still be regarded as highly uncertain, despite these few success stories. Even in New York, private vans carry only a small fraction of bus riders, and the potential for expansion if they were encouraged is questionable. Nevertheless, to the extent that the U.S. market has a large number of low-volume routes, minibuses might be effective. However, the peaked character of U.S. markets may make minibus service too costly for many applications because of high labor costs. The relative strength of the U.S. transit unions also makes it uncertain whether the labor concessions, often critical to instituting a viable minibus operation, could be achieved.

Summary

The basic lesson emerging from these U.S. experiences with transit contracting is that the private sector performs best when it is asked to do a relatively well-defined task, similar to one that a number of competing firms are performing already, and with a minimum of interference by public authorities beyond that required to prevent fraud or other abuses. Conversely, the private sector offers few cost advantages in places where contracting procedures reduce competition or where there is extensive governmental interference in operational decisions.

A natural question, especially given the limited scope and scale of the United States' own experiences with transit privatization, is the extent to which the privatizations and deregulations occurring elsewhere might be successfully applied in the United States. In general, the benefits of full-scale or total transit privatization and deregulation would probably not be as great in the United States as elsewhere. To start, it seems highly unlikely that much of existing bus services would prove commercially viable in the United States. Under the best conditions, only 50 percent or so of existing bus transit in the United States would likely prove commercially viable, mainly inner-city routes; by contrast, more than 80 percent of such services proved commercially viable in Britain after reforms were implemented. Much of the American transit industry would thus likely remain a "ward of government," dependent on subsidies. In such circumstances, the prinicpal benefit of privatization would probably be altering the circumstances for awarding subsidies, that is, converting from intergovernmental grants to bidding processes. These changes would most likely result in significant improvements in cost and productivity but would probably not stimulate as many useful service innovations as elsewhere.

Chapter 6. Lessons from the Global Community

THE BUS TRANSIT INDUSTRY has accommodated a remarkable range of privatization experiences in many different countries at many different stages of economic development, with different policies on competition and regulation, and in different market and economic cultures. Although a common cycle of private beginnings, public takeovers, and reprivatization is observable, countries vary greatly in their experiences. Despite the diversity of environments and policies, however, several common lessons emerge.

The Importance of Competition

The benefits of privatization and deregulation are greatly enhanced if effective competition can be established and maintained. This obvious and most important lesson was apparent among the developing countries; deregulation was more successful among Santiago's shared-ride taxis than its buses, for example, because the shared-ride taxis did not suffer from collusive route associations. In Britain, the metropolitan counties with the most competition generally had the smallest fare increases and largest gains in bus-miles of service (relative to their subsidy cuts). Finally, in the United States contracting out was more likely to generate cost savings, particularly for dial-a-ride services, in places where the services contracted were similar to those private sector firms were already providing and where the bidding and oversight procedures did not discourage entry and competition.

Competition is important for several reasons. In the first place, competition helps evoke the cost savings and service improvements that are usually the key motives for bus privatization. The more competitive the markets, moreover, the more likely bus companies will be to pass cost savings on to passengers or taxpayers (rather than retaining them as higher profits or as a means of financing excessive costs). Furthermore, the greater the cost savings or the innovation stimulated, the more likely there will be something for all to share.

Britain's experience, and to a lesser extent that of the United States, also suggests that the prospects for maintaining or establishing competition are likely to be greater in less developed countries than in the industrialized countries simply because financial viability and prospects of the bus transit industry are less promising when incomes rise and automobile ownership becomes widespread. Thus the transit service or ridership that can be sustained without government subsidy is likely to be smaller in a developed than a developing country. Transit operators can sometimes have a base in a related industry, of course, as in the United States where most of the private firms awarded urban bus service contracts are intercity or charter bus operators or taxi companies. Competition is also not purely a game of numbers; a few firms may provide very effective competition, especially where markets are relatively contestable (that is, where entry and exit are relatively easy). Nevertheless, no market seems perfectly contestable, and small numbers of firms may increase the opportunities for collusion.

Competition may also require some government antitrust enforcement as well, particularly in the developed countries where the market is likely to be intrinsically less competitive. In Britain, for example, a need may exist for vigilance against collusion among the few dominant firms that seem to have emerged in each metropolitan area after deregulation and also against the possibility of predatory behavior by large firms against their smaller rivals. Even in developing countries government intervention may occasionally be needed to maintain competition, as the experience with route associations in Santiago illustrates.

Service Innovations versus Cost Cutting

Privatization can offer important benefits not just in reducing costs and improving productivity but also in encouraging more market-oriented services. The most dramatic example is Britain, where deregulation and privatization stimulated the spread of minibuses, thus replacing less frequent double-deck bus services and extending routes to low-density and previously unserved housing estates.

Innovation is especially important in industries, like bus transit, that are declining because of adverse external trends. Worldwide, personal incomes are growing, which tends in turn to encourage the consumption of higher-quality transportation services and more housing. In the poorest developing countries this increase in income means a shift from foot-powered transportation to buses. In most industrialized countries and a growing number of the wealthier developing countries, these demands are usually accommodated by the spread of the private automobile and suburban housing, thus reducing the normal market for bus transportation. Cutting unit costs can help to hold the market share of buses at least for a while, but further innovations, in quality of service as well as productivity, may be needed.

To the extent innovation occurs, deregulation and privatization might be transformed from policies where some parties benefit, usually taxpayers or consumers, at the expense of others, usually labor, to policies in which there are few or no losers and mostly winners. Labor may suffer some pay cuts or have to work harder, for example, but may also gain from the increased employment opportunities in a larger, leaner, and more market-oriented industry. Consumers may lose from higher fares, particularly if any efficiency gains are not enough to offset the withdrawal of public subsidies or restrictive fare regulations, but the improvement and expansion of services may more than compensate consumers for any fare increases, especially where the past lack of competition or limits on fares or subsidies had severely constrained services and inhibited innovation.

In this context, contracting out is less attractive than more complete forms of privatization. Contracting out can provide incentives to cut unit costs, but since the services are still largely designed by public authorities, contracting offers little opportunity for the private sector to explore service innovations. A privatized, deregulated, and unsubsidized commercial bus sector, by contrast, may be an important font of service innovations.

Achieving a truly commercial bus industry capable of such innovation will be easier in developing than developed countries, however, as bus operators in developed countries have a larger gap to close between farebox revenues and costs than their counterparts in developing countries. Bus ridership is also higher in the developing world, so operators are less handicapped by the diseconomies of serving lightly traveled routes. Nevertheless, some urban bus services hold the potential for becoming commercially viable even in the industrialized countries, and are suitable targets for service innovations, as Great Britain's experiment demonstrated.

The Perils of Regulating Fares or Maintaining Subsidies

It is not easy to maintain a viable and competitive private sector while still regulating fares or maintaining subsidies for unprofitable but socially worthwhile services, especially in developing countries.

Most developing countries have maintained fare regulation during privatization to insure that fares would remain reasonable. Experience suggests, however, that public officials may be tempted to set unrealistically low fares for private buses; these low fares may lead to inadequacies and shortages in services. In several third world cities where the bus industry has recently been privatized but not deregulated, dissatisfaction with inadequate service, usually blamed on the companies rather than on fare regulators, may eventually lead to a renationalization of bus service and a fruitless repetition of the cycle of private and public ownership.

Fare regulation may also discourage service innovation, even if it does not drive the industry into financial difficulty. Regulated fares often tend to be simple or uniform; after all, it seems fair that most riders should pay the same or similar fares. However, simple or uniform fare structures often reflect poorly the costs of various types of service. Fares that do not vary with distance traveled or time of day, for example, may discourage bus companies from expanding service to longer-distance or highly peaked markets. Similarly, uniform regulated fares may discourage or prevent the industry from offering a variety of services (for example, direct express or minibus services at premium fares).

The risk of deregulation, of course, is the possibility that fares will rise sharply. This increase has occurred in several cases, especially when deregulation is accompanied by the withdrawal of substantial subsidies (as in some British metropolitan counties) or when there was little competition (Santiago's buses). If competition is present and any subsidies withdrawn are modest relative to the cost or productivity savings expected from privatization, then the chances of significant fare increases should be sharply reduced.

The British and U.S. experiences suggest that subsidized services, procured from private firms through competitive bidding, may promote rather than inhibit competition by providing a niche for small firms to enter the market. In developing countries, however, the prospects for using subsidies as a tool to enhance competition are weaker, largely because administrative capacities to implement competitive bidding schemes seem more limited. But the need for subsidies is also probably less important in the developing countries as well, largely because their low labor costs and high patronage make it more likely that extensive bus services can be self-supporting.

Furthermore, the experience of developing countries suggests that there are few advantages to a mixed system of subsidized public and unsubsidized private bus firms, largely because the public firms are usually less efficient than their private counterparts, especially when the public and private firms offer service on similar routes and charge the same fares. Even when the public firms charge a much lower fare or provide clearly different services than their private counterparts, subsidies can have significant drawbacks. The most important common problem is that the publicly subsidized services may remain inadequate and over-crowded because of the high costs of the public operator and the continuing inability or unwillingness of the government to foot the subsidy bill for service expansions. Conceivably, too, if the unsubsidized private operators were allowed to provide standard as well as deluxe or premium fare services, they could provide the same or greater capacity at a similar or only slightly higher fare than the public companies. The publicly subsidized alternative may also do more damage (by, say, restraining innovation by privately operated services) than good (by providing a more affordable alternative), especially when the fare differentials are small.

The Limits of the Second Best

Finally, experiences with bus deregulation and privatization in Britain and the developing world cast some doubt on the empirical foundations of an important economic concept—second-best pricing—often used to justify government intervention in a market economy. Normally, economists assert that economic efficiency and welfare are improved if prices in a sector are brought closer to the costs of producing its particular products or services. This conclusion, however, assumes competitive markets elsewhere in the economy, especially for complementary or competitive goods or services. Thus, a move toward more cost-oriented or competitive prices in a certain sector might reduce rather than improve economic welfare if that sector's goods or services are complements or substitutes for other goods and services that are sold in noncompetitive markets (in which prices diverge widely from costs).[1] The term "second best" refers to the environment in which the sector is operating—a first-best environment is one where all substitute or complementary goods and services are produced in competitive markets and properly priced.

1. For classic statements of this proposition see R. Lancaster and R. Lipsey, "The General Theory of the Second Best," *Review of Economic Studies*, vol. 24, no. 63 (1956), pp. 11–32; and Marcel Boiteux, "Marginal Cost Pricing," translated and reprinted in James R. Nelson ed., *Marginal Cost Pricing in Practice* (Prentice Hall, 1964), pp. 51-58.

Second-best principles are often used to justify subsidies to urban bus services. Auto users are usually undercharged for the roads they use or the congestion and pollution they create, particularly for peak-hour commuting in congested urban areas.[2] Subsidies to urban bus transit are needed, it is often argued, to counterbalance the underpricing of auto travel and to insure a more appropriate mix of auto and transit use. Efforts to privatize and deregulate bus services may conflict with these policies, however, especially if transit subsidies are withdrawn at the same time and if nothing is done to correct the underpricing of autos.

The British and developing country experiences suggest, however, that a move to a more market-oriented bus industry and the removal of bus subsidies designed to compensate for the underpricing of auto use do not invariably increase auto traffic congestion or otherwise exacerbate larger urban and environmental problems. Indeed, the bus industry seems to have a much better chance of competing with the automobile if it is lean and market oriented than if it is publicly subsidized but also heavily encumbered by government regulation and constraints. In Britain, minibuses, express services, and other innovations unleashed by deregulation and privatization did not add quite as many transit commuters as reduced subsidies and higher fares subtracted, but most of the lost British bus riders apparently made fewer trips, rather than shifting to autos or other modes. The American experiments with dial-a-ride and express services, though very limited, add further credibility to this hypothesis.

Recent, and fairly realistic, behavioral models of high-density commuter corridor operations also suggest that transit subsidies may do little to reduce the number of autos used in commuting. Rather, transit subsidies may tend to pull commuters away from walking and car pooling.[3] In essence, the cross elasticities seem greater between walking and transit, and carpooling and transit, than between driving and transit. As a consequence, higher transit subsidies may not so much reduce the number of cars on the road in peak hours as reduce the number of people walking to work or riding in carpools. Another consequence of transit subsidies, as Britain's experience suggests, may be to induce people to take trips that they otherwise would not.

Skepticism about the efficacy of second-best policies is strengthened if, as seems likely, the shadow price on each unit of government spending is

2. See, for example, John R. Meyer and Jóse A. Gómez-Ibáñez, *Autos, Transit and Cities* (Harvard University Press, 1981), pp. 188–208, 290–95.

3. See, for example, Paul D. Kerin, *Efficient Transit Management Strategies and Public Policies: Radial Commuter Arteries*, Ph.D. dissertation, Harvard University, May 1990.

greater than one. An additional $1.00 in taxation may create losses elsewhere in the economy of more than $1.00 because of distortions caused by the additional tax, or by collection or other administrative costs. If so, the benefits of implementing a second-best pricing scheme that involves government subsidy, as most such schemes do, must be correspondingly greater to render a net improvement in welfare. Estimates of the shadow price to be attached to an extra unit of taxation range upward to nearly $1.60 or so; $1.40 might not be too far from the consensus.[4] These estimates mainly pertain to industrialized countries (for example, the United States, Britain, and Australia), but it seems unlikely that they are markedly lower in developing countries.

Public transit subsidies and operations are sometimes alleged to reduce traffic congestion and improve traffic safety in other ways. For example, public provision may reduce the problem of aggressive driving by profit-motivated private bus drivers. Reports of aggressive driving are fairly common in cities of developing countries, especially when many small private firms or individual owner operators ply the same route. Some observers argue that better public provisions for on-street bus stops, off-street terminals, and the enforcement of traffic regulations would ease these problems. In some cities the problem has been relieved by the development of route associations, which set schedules and sometimes share revenues among the independent operators on a given route; but route associations, as noted in chapter 3, may also act to limit competition in undesirable ways and may even create excessive bus congestion (as in Santiago, Chile).

Privatization is sometimes blamed for increasing traffic congestion by promoting the use of smaller vehicles, such as minibuses, in lieu of full-sized buses. This criticism may be misguided in some of the wealthier developing or less wealthy developed countries, where passengers deprived of minibuses or jitneys might select an autobike, or even an auto, rather than a conventional-sized bus for their commutation if forced to choose. As noted earlier, moreover, there is no inherent reason why private operators could not be required to use larger vehicles. Private operators tend to select smaller vehicles when given the opportunity, however, because they are cheaper to operate and are usually popular with customers. The cost and service advantages of smaller vehicles are

4. E. Browning, "The Marginal Cost of Public Funds," *Journal of Political Economy*, vol. 84 (April 1976), pp. 283–98; and Dale W. Jorgenson and Kun-Young Yun, "The Excess Burden of Taxation in the United States," *Journal of Accounting, Auditing, and Finance*, vol. 6 (Fall 1991), pp. 487–509.

obviously important considerations from the public's perspective and should be balanced against the greater demands for street space in deciding whether vehicle sizes should be restricted.

Overall Assessment

On the whole, urban bus services seem to be a highly promising area for privatization. In its least ambitious form—contracting out of services—privatization seems to reduce costs by at least 20 percent and often much more than that. Furthermore, contracting out to private firms sometimes can stimulate similar cost reductions in any remaining public operations, particularly if the public operators believe that contracting will expand if their performance does not improve.

More ambitious forms of privatization offer even more potential but also some additional risks. Privatization without subsidies but with continued government regulation of fares seems to have worked fairly well where it has been tried (mainly in the cities of developing countries). This form of privatization should evoke service innovations as well as productivity or cost improvements. The principal risk is that the retention of public regulations, especially for fares and entry to the industry, might eventually seriously limit the ability of private operators to maintain or expand services and thus possibly lead to renewed calls for public takeovers and subsidies.

The most ambitious reform is to combine privatization with deregulation, particularly of fares. Under this regime the scope for service innovation is greatest, since companies can experiment with various combinations of fares as well as services. The risk that regulation will lead once again to bankruptcy and a return to public ownership is also reduced, enhancing the scope for investment and experimentation. This policy combination is slightly more dependent than the others on the maintenance of effective competition, however, since without competition unwarranted and politically embarrassing fare increases may occur.

The conclusion that the more ambitious forms of bus privatization are often the most promising holds more strongly in developing than developed countries. In the developing world privatization frequently generates large expansions in service with little or no fare increases, even without deregulation and even when public subsidies to bus operators are simultaneously withdrawn or reduced; as a consequence, riders almost surely gain and labor has even more job opportunities to compensate for any lowering of wage rates. This happy outcome seems largely to have come about because privatization has evoked larger unit cost reductions in developing than developed countries, owing to a combination of the greater competition among private bus operators and, in some cases, the greater initial inefficiency of the public companies.

Deregulation may offer somewhat greater benefits in the developing countries than elsewhere, although the experience is limited. In the few places where fares have been deregulated in the developing countries, fare increases were rather modest (certainly compared with increases in Britain). Continuing regulation often risks inadequate services, moreover, since regulators in the developing world are under great pressure to set unrealistically low fares. In short, in developing countries, deregulation of fares probably offers less risk of excessive fare increases and more gains in improved or more adequate service than in the industrialized countries.

As a consequence of all these factors, the desirability of maintaining public subsidies in a privatized and deregulated urban bus industry may also be less important in the developing countries than elsewhere. If privatization and deregulation by themselves offer the prospect of expanded services with only modest or no fare increases, the risks of continuing subsidies may not seem worthwhile. Accordingly, privatization and deregulation are probably desirable public policies, taken separately or together, for developed and developing countries, but this advice can be rendered with fewer qualifications in the less developed parts of the world than in the developed.

Ultimately, though, the public policy contribution of the more ambitious and complete reforms, such as those in Britain and a few of the less developed countries (for example, Sri Lanka and Chile), may reside in their long-run implications for rationalizing choices about the future structure and development of urban areas. Fundamentally, complete deregulation and privatization forces a restructuring of the transit industry in which costs are better controlled, fares are more closely related to costs, and services are tailored to identifiable market demands. If accompanied by a rationalization of user charges imposed on automobiles, such that the full congestion and parking costs of using automobiles were assessed, then the stage would be set for society to make reasonably sensible choices about the extent to which it preferred high-density metropolitan development supported by public transit rather than lower-density development dependent on automobile commuting. In short, thorough transit reforms, as in Britain, represent one large step away from the inefficiencies of subsidies to autos and public transport, as widely practiced today. These subsidies, often costly in direct monetary terms, tend to waste resources and distort society's choices about how and where to live and work. Society might be better served by breaking the cycle of transit privatization-nationalization and regulation, letting prices approach costs for all urban transport modes, and thus creating a more rational basis for choices about urban structure.

Part Two

Privatizing Infrastructure: High-Performance Highways

Chapter 7. Highways: Trends and Issues

HIGHWAYS ARE ARGUABLY the most important and costly form of transportation infrastructure in developed and developing countries. Most domestic passenger and freight traffic moves by highways in most market economies, and highway traffic is increasing faster everywhere than most other modes. Not surprisingly, therefore, many countries are experimenting with highway privatization, especially with private toll roads.[1]

Toll Roads in Developed and Developing Countries

Most developed countries have built extensive high-performance, access-controlled, and grade-separated expressway systems since the end of World War II. These countries can be classified into two groups: those that relied primarily on motor fuel or general tax revenues to finance their expressways and those that relied heavily on toll receipts.[2] Tax financing predominates in northern Europe, the United States, Canada, and Australia. By contrast, toll financing predominates in southern Europe and Japan.

1. Private toll roads are the focus because most major private roads are financed largely by collecting tolls from highway users. The principal exceptions are local roads needed for land development projects from which a private company hopes to profit. In many countries, a private company developing a large real estate project, such as a residential subdivision or an office park, will often build the roads within its project without resorting to toll financing. Private land developers may build or contribute to the cost of roads nearby but off site, often in return for government planning permissions for their project. Once built, however, these on- and off-site roads are usually owned, maintained, and operated by public authorities rather than the land development company. By contrast, for nonlocal high-performance highways, where the profits from related land developments are often widely diffused and not easily captured, a private road will almost invariably be tolled.

2. For an overview of road financing systems in developed countries see Organization for Economic Cooperation and Development, *Toll Financing and Private Sector Involvement in Road Infrastructure Development: Report* (Paris: OECD, 1987), esp. pp. 83–107.

Though one form of financing usually predominates, most developed countries have used both systems of finance to some degree. Thus, while tax financing is the norm in the United States, approximately 4,000 miles of the 55,000 miles of U.S. expressways are tolled, as are some high-cost bridges and tunnels. Most of these toll expressways were built before 1956, when Congress decided to fund the construction of a 43,000-mile Interstate and Defense Highway System primarily financed by federal gas tax receipts. Similarly, the countries that rely heavily on tolls do not toll every high-performance highway. In France, Spain, and Italy, urban or commuter expressways are seldom tolled even though intercity expressways usually are. Almost every country that uses tolls also requires that a parallel untolled route be available to motorists, even though the alternative is usually not built to expressway standards.

Many of the countries that have traditionally relied on tax financing became increasingly interested in tolls during the 1980s, usually for financial reasons. U.S. interest in toll roads revived in the mid-1980s, for example, after the interstate system was largely complete and inflation had eroded the purchasing power of federal and state motor fuel taxes.

Even more striking is the growing acceptance of private toll road concessions as an alternative to public toll roads. Several southern European countries have had decades of experience with private toll road companies, especially Spain and France. Other developed countries are now experimenting with private toll roads for the first time. Britain granted two concessions for toll bridges to private companies in the late 1980s, for example, and awarded its first toll road concession (for a thirty-mile, six-lane bypass around the northeast corner of Birmingham) in 1991.[3] In the United States, two states, California and Virginia, signed franchise agreements with private companies for five toll roads by the early 1990s, and several other states have passed legislation authorizing private roads.[4] Neither the British nor the U.S. private highways had opened as of late-1993, although one is under construction.

Most developing countries have just started to build high-performance expressways in the last two decades, and many are now relying heavily on

3. The two bridges are the Dartford and the Severn bridges. For a description of the Birmingham road see Peter Reina,"Long-Term Profits Lure Trafalgar House into High-Risk BOT Bypass for Birmingham," *Public Works Financing* (October 1991), pp. 21–23

4. For a description of the U.S. experience see José A. Gómez-Ibáñez and John R. Meyer, *Private Toll Roads in the United States: The Early Experience of California and Virginia*, report to the University Transportation Center, Region One (Harvard University Taubman Center for State and Local Government, December 1991) .

toll financing and private concessions.[5] Typically, they have begun by building expressways in the heavily traveled corridors near or in their largest cities. The earliest expressways were usually built and operated by government-owned toll road companies. Beginning in the mid-1980s, however, many developing countries began to grant new toll road concessions to private rather than public firms and undertook the building of more ambitious intercity networks.

Mexico is a fairly advanced but otherwise typical example. Beginning in 1963, a government-owned toll road company slowly built almost 1,000 kilometers of tolled expressways, mainly near Mexico City. In the late 1980s, however, the government began to award private concessions for new toll roads to increase the rate of expressway construction. This effort accelerated when President Carlos Salinas took office in 1989 and announced a plan to build an extensive national expressway system by granting 5,300 kilometers of new toll road concessions between 1989 and 1994. Over 1,000 kilometers of private toll roads were already in service by 1992, with more under construction.[6] In Southeast Asia, Thailand, Malaysia, and Indonesia are following a similar, if slightly less ambitious, pattern.[7]

Since the dissolution of the Eastern bloc in the late 1980s, Eastern Europe has also increasingly considered toll roads and private concessions. Toll road authorities owned by regional governments had been building a tolled trans-Yugoslavia highway before the breakup of their country; in late 1991, the Serbian government signed a concession with a consortium led by an Italian company to build and operate a sixty-mile expressway segment.[8] Czechoslovakia is planning a concession from

5. For an overview of the developing country experience see Frida Johansen, ed., *Earmarking, Road Funds and Toll Roads: A World Bank Symposium*, Report INU-45 (Washington: World Bank, June 1989) .

6. Robert N. Panfield, "Toll Highways in Mexico," pp. 157–70, in Johansen, *Earmarking, Road Funds and Toll Roads*; Gabriel Castaneda Gallardo, "Financing, Construction and Operation of Highways through Concessionairy Arrangements in Mexico," in Johansen, ed., *Earmarking, Road Funds and Toll Roads*, pp. 171–96; Victor M. Mahbub M., "Mexico's Private Road Concessions," *Public Works Financing International*, June 1991, pp. 1–5; and Mitchel Stanfield, "Modernizing Highway Infrastructure through Toll Concessions in Mexico," report to the U.S. Department of Transportation, Federal Highway Administration (Washington, December 1991).

7. Frida Johansen, "Toll Road Characteristics and Toll Road Experience in Selected South East Asian Countries," *Transportation Research A*, vol. 23A, no. 6 (1989), pp. 463–66; Maurice Le Blance, "Toll Road Experience in Malaysia," pp. 142–47 in Johansen, *Earmarking, Road Funds and Toll Roads*.

8. N. Cengiz Yucel, "Toll Roads and the Financing of Road Expenditures in Yugoslavia," pp. 148–56 in Johansen, *Earmarking, Road Funds and Toll Roads*; and "Serbian Road Concession," *Public Works Financing*, January 1992, p. 17.

Prague toward Germany, and Hungary is selecting international private consortia to bid on its first planned forty-three-kilometer toll road concession.[9]

Key Issues

Some of the issues raised by private toll roads are similar to those raised in the debate over private provision of bus services. Proponents contend that the private operators can build or operate a service more efficiently, for example, or will be more innovative in the facilities they design and services they offer. Similarly, the extent to which privatization transfers costs and benefits among parties is a consideration in toll roads as it is in buses.

Private toll roads raise other issues, however, that are of lesser importance or not found in the debates over bus privatization. These issues arise primarily because roads are expensive, capital-intensive, long-lived, and immobile; bus services, by contrast, are relatively more labor intensive and the capital (primarily vehicles) is mobile and comparatively short-lived. Most important among these additional highway-related issues are (1) the possibility that privatization will increase capital investment in such facilities; (2) the merits of toll versus tax financing; (3) problems of regulating long-lived private concessions, which often have monopoly possibilities; and (4) the emergence of environmental and other opposition to siting major facilities.

The Potential for Increasing Capital Investment

For toll roads and other types of infrastructure the possibility of attracting increased capital investment is often as important a motivation for privatization as the potential for improved efficiency. High-performance highways are expensive, and most governments are hard-pressed to maintain and improve their roads in the face of rapid growth in highway traffic. Private provision of roads, advocates argue, will increase the funds available for highway investment by tapping private capital markets. If the private markets finance roads, governments can devote their scarce tax resources to other pressing social needs and services where private markets or finance are less workable or appropriate.

9. "Czech Toll Concession Interest" and "Klienwort Wins Hungarian Plum," *Public Works Financing*, January 1992, pp. 20, 16; and "East Europe's First BOT Road," *Public Works Financing*, November 1991, p. 6.

This argument has two important limitations. In the first place, private financing of roads will probably come at the expense of other forms of private investment, since privatization usually does nothing (at least directly) to increase the pool of savings on which private capital markets draw. This limitation may be less serious in developing than developed countries because private toll roads in developing countries can sometimes tap international capital markets that are otherwise not available to such nations. But in countries where private toll roads draw on the same capital markets as other domestic private investors, privatization is unlikely to increase total investment in the society. In such cases, the economic rationale for private provision of roads fundamentally rests on investments in roads proving more productive than the investments they displace.

A second limitation is that privatization may not be necessary to tap private capital markets. Government road authorities can draw on private capital markets by issuing bonds backed by toll or tax receipts, just as private companies can. In short, the case that private provision increases highway investment is strongest only where government toll road authorities are for some reason denied access to private capital markets. Even then, private investment in highways will usually come at the expense of other forms of domestic private investment unless private road companies have access to capital markets that are otherwise closed.

The Merits of Toll Financing

Private toll roads seldom rely entirely on toll receipts to pay all their costs. They often receive government aid in the form of donated right-of-way, for example, or in cash contributions to defray part of their construction expenses. Nevertheless, toll receipts are usually by far their primary source of revenue. Indeed, since the primary motivation for road privatization is usually the shortage of tax resources, government enthusiasm for private roads increases greatly if most expenses can be financed out of tolls.

Transportation planners and economists have long debated the wisdom of financing roads from tolls rather than taxes, irrespective of whether these roads will be privately or publicly built and operated. One basic concern is whether toll receipts should or will cover costs. It is not always possible to recover the costs of building and maintaining an expressway through tolls where traffic volumes are low or costs unusually high. Low traffic volumes are a problem in developing countries; they also occur in developed countries where the expressway network has been extended to ever more lightly traveled routes. Even so, tolls can be viewed as a useful supplement to limited tax resources by financially constrained governments.

Another key issue is whether tolls are likely to encourage efficient or responsible highway use and investment, especially when compared to the alternative of financing highways exclusively from fuel, motor vehicle, or broad-based taxes. Advocates argue that if tolls are set to cover the marginal or incremental costs of serving different types of vehicles on a facility, motorists should be encouraged to use the facility only when their benefits from doing so are at least as great as the costs of their road use. Toll receipts may provide a useful guide to the merits of additional highway investment as well, since most benefits of highway improvements accrue to motorists, and tolls reflect the value motorists place on the services they receive.

Tolling does not necessarily improve highway investment and use, however. The argument that it will encourage efficient use usually assumes that tolls are set at or reasonably close to marginal or incremental costs (including congestion and other spillovers); these costs typically vary by road type, location, and time of day. Simpler or less accurate pricing schemes may be adopted in practice, however, to reduce collection costs or increase political support. Incremental cost pricing may also conflict with the objective of recovering costs, particularly if a toll road has fairly low traffic volumes relative to its design capacity.

That toll roads usually compete to some degree with parallel untolled highways can also create second-best pricing problems similar to those found in buses (and discussed in chapter 6). In this case, tolling one road but not the other may cause excessive congestion on the untolled facility and underutilization of the toll road. Under most conditions, the optimal (first-best) solution would mean tolling both facilities, with the differential toll rates reflecting the differential in social costs of using the two facilities. Where tolling both facilities is politically unacceptable or administratively impractical, it may be desirable to lower tariffs on the toll road to correct the misallocation of traffic.[10] These toll reductions may, in

10. The seminal statement of this problem can be found in A. C. Pigou, *The Economics of Welfare* (London: Macmillan, 1920). For further discussion of the "two-road problem" see John R. Meyer and Mahlon Straszheim, *Pricing and Project Evaluation*, vol. 1 of *Techniques of Transport Planning*, ed. John R. Meyer (Brookings, 1971), pp. 44–59 . For a discussion of the practical impact of this problem in Britain see Kenneth J. Button, "Impact of Toll Policy in the United Kingdom," *Transportation Research Record*, no. 1107 (1987), pp. 55–64.

turn, reduce the financial viability of the toll road as a stand-alone enterprise.

Finally, collecting tolls requires toll plazas and equipment, collection staffs, and delays to motorists. Although these costs may be reduced by careful design of the toll collection system and use of modern electronic collection technologies, they are never eliminated and thus constitute a cost burden on toll facilities, to be offset, it is hoped, by achievement of greater efficiencies elsewhere. Of course there also are costs of collecting taxes, in collection staffs required and in the distortions these taxes may cause in the economy.

In general, the advantages and disadvantages of tolls must be weighed against those of alternative financing schemes or of building fewer roads. Toll financing may not encourage optimal road use and investment, for example, but financing through fuel or broad-based taxes may do worse or no better.

The Problem of Economic Regulation

Another problem that arises in the road privatization debate is the possibility that a private company may be tempted to exploit any monopoly power it might enjoy by raising rates well above costs or by constraining capacity. As already observed, this problem can occur with buses as well, especially in developed countries where the market for bus services may be thin and competition difficult to sustain. Competition in the bus industry is easier to create, however, if only because the minimum capital investment required to begin service is not so enormous and because vehicles can be shifted among routes or cities as competitive opportunities arise. Roads, by contrast, are far more costly and immobile and almost invariably require substantial land assembly and environmental clearances. In short, the barriers to entry are simply much greater with highways than with buses.

Most private toll roads usually do face competition, of course, from public untolled roads. These free alternative roads are usually not built to expressway standards, however, and are usually more circuitous than toll roads or pass through the centers of small cities and towns where speeds are restricted. Indeed, where free roads compete, a toll road must have some market power or competitive advantage to be able to charge tolls; otherwise all motorists would use the free road.

Whether the market power enjoyed by a toll road is sufficient to justify the problems inherent in government regulation is an open question. The answer almost surely varies from case to case. Most countries with toll roads feel reluctant to rely entirely on the market (free road competition) to discipline the behavior of private toll road operators, however, and thus have moved to regulate private toll rates or the allowed rates of return on investment, or both.

Economic regulation is not easily done well, though, as the experience with buses demonstrates. The central problem is to strike a balance between protecting the public from potential abuses of monopoly while insuring that private enterprise has an opportunity to earn an adequate return on its investment. This balancing act is more complicated for toll roads than for buses, moreover, because of the greater cost, lesser mobility, and longer life of road investments. A road may not recoup its investment for twenty or thirty years. Private road investors therefore may need assurances that the regulatory regime or rules will be relatively stable or predictable over the life of their concession before they are willing to invest. For its part, the government needs long-term assurances too, that the road will be operated and maintained well, and that it will be in good condition when it is turned over to the government at the end of the concession period. The desire of both parties for some stability or predictability also must be balanced against the inevitability of changing circumstances over the long lives of these projects. Either party may wish to modify the regulatory or concession agreement in the face of unforeseen developments.

Special regulatory problems also arise (as in buses) when government assistance is given to private companies. Government assistance or guarantees may be necessary to attract private capital, for example, if the toll road is considered a highly risky or speculative venture. But a company with too little of its own resources at risk may not always behave responsibly.

In short, public authorities and private companies must work out reasonable agreements on whether or how tolls will be regulated, how an adequate return on investment will be determined, and whether and how these rules might change over time. These agreements may vary from road to road and country to country, moreover, since the business risks, regulatory traditions, and governmental administrative capacities also vary.

Siting and Environmental Controversies

Siting is obviously a much more important problem for roads than for buses. Communities along a proposed bus route usually welcome the services provided, although they may occasionally object to the noise or traffic congestion sometimes created, especially if the buses are serving only through traffic rather than local traffic. By contrast, a road requires the taking of local land, often through the powers of eminent domain. Moreover, a new high-performance highway is often perceived by neighboring communities and environmentalists to offer far fewer benefits than costs. It may pass through environmentally sensitive habitats, for example, and the growth and land development it may induce are not always welcome.

Highway siting and environmental controversies have been most common and acute in developed countries, especially in recent decades. For example, some observers argue that the slowdown in new highway construction in the United States since the 1970s has occurred because of increasing controversies over siting rather than growing constraints on government budgets.

Private road companies may have some advantages in coping with such issues. They often have more flexibility to negotiate compensation or mitigation arrangements with opponents than a government agency would, and they have more incentive to do so because they are more sensitive to the financial costs of construction delays. However, community and environmental activists, rightly or wrongly, may fear that a private company is less likely to live up to its environmental obligations than a public agency. Moreover, the sensitivity of private companies to delays may work to their disadvantage as well, since they may shy away from or abandon projects with high potential for controversy.

The uncertainty about these advantages and disadvantages has led to different policies about the degree to which governments or the private toll road companies assume the environmental or siting risk. Usually, the government assumes most of these risks and takes responsibility for picking the routes, securing the necessary environmental permits, and delivering title to the right-of-way. But in several cases, especially in the United States, the private toll road company has been asked to assume many of these responsibilities and risks.

Economic Development and Highway Privatization

Obviously, the potential for private provision of toll roads varies greatly according to the characteristics and environments of the toll road projects. The various countries that have experimented with private toll roads have also adopted different strategies for their promotion and regulation and have varied these policies over time as they have accumulated experience. As a result, much can be learned by contrasting the experiences of Europe, the United States, and the developing countries.

And these experiences do contrast sharply. To start, the stage of economic development is crucial in conditioning the acceptability of highway privatization. Strikingly, highway privatization probably faces the most difficulty at the two extremes of economic development. At the two extremes, it is more difficult for private toll road undertakings to achieve the financial viability that is, almost by definition, normally essential to privatization's success.

At one extreme, in very underdeveloped countries just beginning to industrialize and to create a consumer society, there is simply little demand for high-performance highways. Households own few private cars, and even trucks are scarce especially outside of the largest metropolitan areas. Thus, even though a new highway may face little competition, it probably cannot be financed by tolls but will require some form of tax-financed subsidy to be viable.

By contrast, in a highly developed economy, with high income levels and widespread auto ownership, most of the obviously needed links in the highway system will have already been built. Accordingly, it is difficult to find new links that private vendors could build and toll sufficiently to cover all costs and provide an adequate return on capital. The highway linkages left to be done when an economy is at an advanced stage of economic development are generally of two types: those extending into new development areas with an expectation of initially experiencing relatively low densities of traffic; or very high-cost linkages needed to relieve urban peak-hour congestion. The low-density or development roads extending outward from an already well-established system may not be too expensive to build but may take years to achieve sufficient toll revenues. Congestion-relievers are usually not only expensive to build but also have to recoup their returns over just a few hours of a day, since at off-peak hours the existing highway system is often more than adequate.

The best prospects for highway privatization are thus likely to occur between the two extremes of economic development, and particularly at an early stage of acceleration into a modern industrial or consumer society. Highway privatization seems to work best when automobile ownership first becomes prevalent in a society where high-performance highways are still virtually nonexistent. Under such circumstances, it is easy to identify where a new facility can be constructed and quickly placed on a self-sustaining financial basis.

Chapter 8. The Modern Pioneers: France and Spain

FRANCE AND SPAIN HAVE ARGUABLY had the longest and most extensive experience of building modern, private, high-performance roads financed by tolls in the world (although Italy is also a contender).[1] The Spanish private concessions began in the 1960s, while their French counterparts date from the early 1970s. Both countries are classified as developed but vary in important ways that affect their highway traffic and networks. France has much higher population densities and somewhat higher income levels than Spain, and thus higher average volumes of traffic on its expressways. The two countries have also adopted somewhat different policies toward their private concessionaires over the years, making their comparative experiences highly informative.

The Development of the French Autoroutes

France has experimented with toll and nontoll financing as well as with publicly and privately owned toll road companies in building its system of high-performance expressways, known as autoroutes. By the early 1990s, France had opened approximately 6,000 kilometers of intercity and 1,500 kilometers of urban autoroute (table 8-1). All but 500 of the 6,000 kilometers of intercity autoroutes are tolled; the exceptions are in Brittany and Lorraine, where some untolled autoroutes were built in the hopes of stimulating regional growth, and on several intercity autoroutes where no satisfactory untolled alternative is available. The 1,500 kilometers of urban autoroutes are all untolled, although the

1. Italy's experience, however, by several measures involves little "pure" participation by the private sector. More than 80 percent of the Italian autostradas were built by companies controlled by the national or local governments. Only the early Torino-Milan road was a purely private endeavor. Interview with Franco Schepis, vice president, AUTOSTRADA (Italy), October 1, 1991, in New Orleans.

TABLE 8-1. Kilometers of Autoroute in France, Selected Years, 1955–91

Year	Urban (nontoll)	Intercity autoroutes: Concession (toll)	Intercity autoroutes: Nonconcession (nontoll)	Subtotal	Grand total
1955	77	0	0	0	77
1960	164	10	0	10	174
1965	279	341	33	374	653
1970	474	1,059	66	1,125	1,599
1975	861	2,266	274	2,540	3,401
1980	1,181	3,707	347	4,054	5,235
1985	1,407	4,589	475	5,064	6,471
1990	1,424	5,496	496	5,992	7,416
1991	n.a.	5,728	n.a.	n.a.	n.a.

SOURCES: Figures for 1990 and for concessions from 1955 to 1991 supplied by Bruno Bieder, Highways Directorate, Ministry of Equipment, Paris, letter to the author; other figures for 1955, 1960, 1970, 1980, 1985, from Bernard Seligmann, "General Presentation of the French Highway System," in *France*, vol. 2, *Toll Roads: Report of a Seminar cum Study Tour* (Bangkok: United Nations, Economic and Social Commission for Asia and the Pacific, 1985), pp. 22–23, 48. Figures for 1965 and 1975 from Alain Fayard, *Les Autoroutes et Leur Financement, Notes et Etudes Documentaires*, nos. 4597–98 (Paris: La Documentation Française, 1980), pp. 48–49.

French government recently reversed its long-standing policy against urban tolls by approving a few new tolled urban autoroutes. Besides its autoroute system, the national government maintains a network of 29,700 kilometers of untolled "national roads," including approximately 3,800 kilometers built with two lanes in each direction and separated by a median but designed to less stringent standards than those of an autoroute.

The tolled autoroutes are operated as concessions by eight major toll road companies and two smaller tunnel companies. Only one of the existing major concessionaires is privately owned (Cofiroute, which operates 732 kilometers of autoroute). During the 1970s there were four privately owned companies, but three of them eventually went bankrupt and were taken over by the public sector. In the early 1990s, however, several other private companies began negotiating with national and local governments for new toll road concessions. The remaining seven major autoroute companies and both tunnel companies are sociétés d'économie mixte (SEMs), or mixed investment companies, in which both private and public ownership (in theory) can be combined. By law, public or governmental bodies must own a predominant interest in the SEMs, and in practice these companies are owned almost entirely by the national and local governments.

Origins of the First Public Toll Road Companies, 1955–69

France's autoroute system emerged in response to the combination of rapid traffic growth and constraints on the government budget during the decade after World War II.[2] The government's first major program for postwar road improvements, formulated in the early 1950s, was to have been financed by dedicating a share of the gasoline tax receipts to road construction. Competition for these tax resources prevented the road program from being fully funded, however, while continued traffic growth made the initial plans seem inadequate.

In response, the government passed a law in 1955 to allow toll financing of autoroutes under special conditions. "The use of autoroutes is, in principle, free," the law stated, and tolls are to be used only "in certain exceptional circumstances." Public control over tolled autoroutes would be maintained, moreover, by granting concessions only to "a local public organization, a group of local organizations, or to a chamber of commerce, or to a society of mixed economy in which public interests are upheld." Tolls could be used only for the repayment of the costs of building the autoroute, for its maintenance, and for its eventual expansion, and the concession agreement, including toll rates, had to be approved by the Council of State after consultation with local authorities.[3] The concessions could last up to thirty-five years, at which time they would revert to the state.

In keeping with the spirit of the 1955 law, the government moved cautiously in the late 1950s, creating only two autoroute SEMs and granting them relatively short initial concessions. The government's slow progress stimulated renewed debate over the autoroute program even before the first two SEMs began construction.[4] Motoring and roadway

2. This early history is based largely on Alain Fayard, *Les Autoroutes et Leur Financement, Notes et Études Documentaires*, nos. 4597–4598 (Paris: La Documentation Française, 1980), pp. 25–40; and on personal interviews with Christian Leyrit, general director, Highways Directorate, Ministry of Equipment, Housing and Spatial Development, on October 28, 1991; Henri Cyna, president of Cofiroute, Paris, on September 30, 1991, and October 30, 1991; Robert Lafont, délégué général of Association des Sociétés Françaises d'Autoroutes, Paris, on October 1 and 30, 1991; and Daniel Tennenbaum, president of SANEF, Paris, on October 1 and 29, 1991. For a history of the French system in English see José A. Gómez-Ibáñez and John R. Meyer, "Toll Roads and Private Concessions in France and Spain," draft report to the Department of Transportation, Harvard University, Taubman Center for State and Local Government, February 1992, pp. 9–48.

3. April 18, 1955 law (no. 55-435) as quoted in Fayard, *Les Autoroutes*.

4. For an account of this debate see Fayard, *Les Autoroutes*, pp. 30–33 .

construction interests, led by the Roadway Union of France, argued that the initial concessions were pathetically inadequate, especially since the government had built only seventy-seven kilometers of autoroute (all untolled and in urban areas) by 1955. The debate was complicated by a counterproposal, advanced by the French geographer Jean-François Gravier, for a national network of four-lane express roads built to less stringent standards than autoroutes. Express roads would cost less per kilometer to build, Gravier argued, and thus permit a larger and truly national network of high-performance roads that would better encourage the development of more isolated and lagging areas of the country. By contrast, autoroutes were so costly that only the most heavily traveled corridors could be developed, especially if tolls were to be the primary source of finance.

The debate was resolved in 1960 when an interministerial task force recommended a fifteen-year plan to build 3,500 kilometers of tolled intercity autoroutes.[5] Unlike Italy and other neighboring countries, France already had a national network of relatively high-quality, two-lane roads, the task force argued, and congestion problems were confined to only a small part of that network. The express road design would be inadequate for the traffic volumes projected on some segments, so autoroute construction would be inevitable. Potential profitability should determine construction priorities for toll autoroutes, since autoroutes were only needed where traffic was heaviest. Profitability would also relieve the potential burden on the government treasury. In keeping with the spirit of the new plan, a 1960 decree deleted the reference to "exceptional circumstances" from the 1955 autoroute law. The exception was now the rule, at least for intercity autoroutes.

Five SEMs were created to carry out the 1960 plan, including the two that had been established in the late 1950s (table 8-2). Three SEMs (SANEF, SAPRR, ASF) were to build the main north-south route of Lille-Paris-Lyon-Marseille, which links France's four most important cities, carries by far the heaviest traffic volumes, and was the highest priority (figure 8-1).[6] The other two SEMs (ESCOTA and SAPN) were to build two important east-west routes: from Nice to Marseille along the Côte

5. This 1960 autoroute master plan and the three subsequent autoroute master plans issued irregularly since they deal only with intercity autoroutes. Nontoll autoroutes and national roads are delineated in a series of five-year master plans, the first being the 1952 plan described earlier; Leyrit, interview.

6. The 1960 plan called for Lille-Paris-Lyon as the top priority and the priorities were formally adopted and extended to Marseille in a 1963 decree; Cyna, interview.

TABLE 8-2. The Five Original French SEMs

Company	*Intended route*	*Date created*	*Law passed*	*Initial equity (millions of francs)*
Société de l'Autoroute Estérel–Côte d'Azur (ESCOTA)	Nice–Marseille	1956	1957	15.0
Société de l'Autoroute de la Vallée du Rhône (SAVR), later renamed Société des Autoroutes du Sud de la France (ASF)	Marseille–Lyon	1957	1961	2.0
Société de l'Autoroute Paris–Lyon (SAPL), later renamed Société des Autoroutes Paris–Rhine–Rhône (SAPRR)	Lyon–Paris	1961	1963	0.85
Société des autoroutes du Nord de la France (SANF), later renamed Société des autoroutes du Nord et de l'Est de la France (SANEF)	Paris–Lille	1963	1963	0.5
Société de l'Autoroute Paris–Normandie (SAPN)	Paris–Caen	1963	1963	0.5

SOURCE: Fayard, *Les Autoroutes et Leur Financement*, pp. 25–40.

d'Azur (a route crowded with tourists in the summer) and from Paris west to Caen and Le Havre.

All five SEMs were capitalized with relatively nominal equity, largely because the majority shareholders were local communities and chambers of commerce along the routes with few resources to commit to road construction. The national government had a small share of the equity in each SEM (typically 20 percent), held by the Caisse des Dépôts et Consignations (CDC), a powerful French national investment bank.[7] By law, delegates of the national government served on the SEMs' boards of directors.

The national government recognized that it would have to give the SEMs some initial assistance, in part because of their limited equity. Accordingly, a law passed in 1958 allowed the state to guarantee the loans of the SEMs and to provide cash advances to cover a percentage of the initial construction costs as well as in-kind advances (advances en nature) of right-of-way or roadways already owned or built by the state. Cash "balancing" advances could be provided to cover early-year operat-

7. An exception was ESCOTA, in which the Caiise des Dépôts et Consignations (CDC) had a 45 percent share and a CDC subsidiary, SECT, held additional stock; see Fayard, *Les Autoroutes*, pp. 145–52.

FIGURE 8-1 France's 1960 Autoroute Masterplan

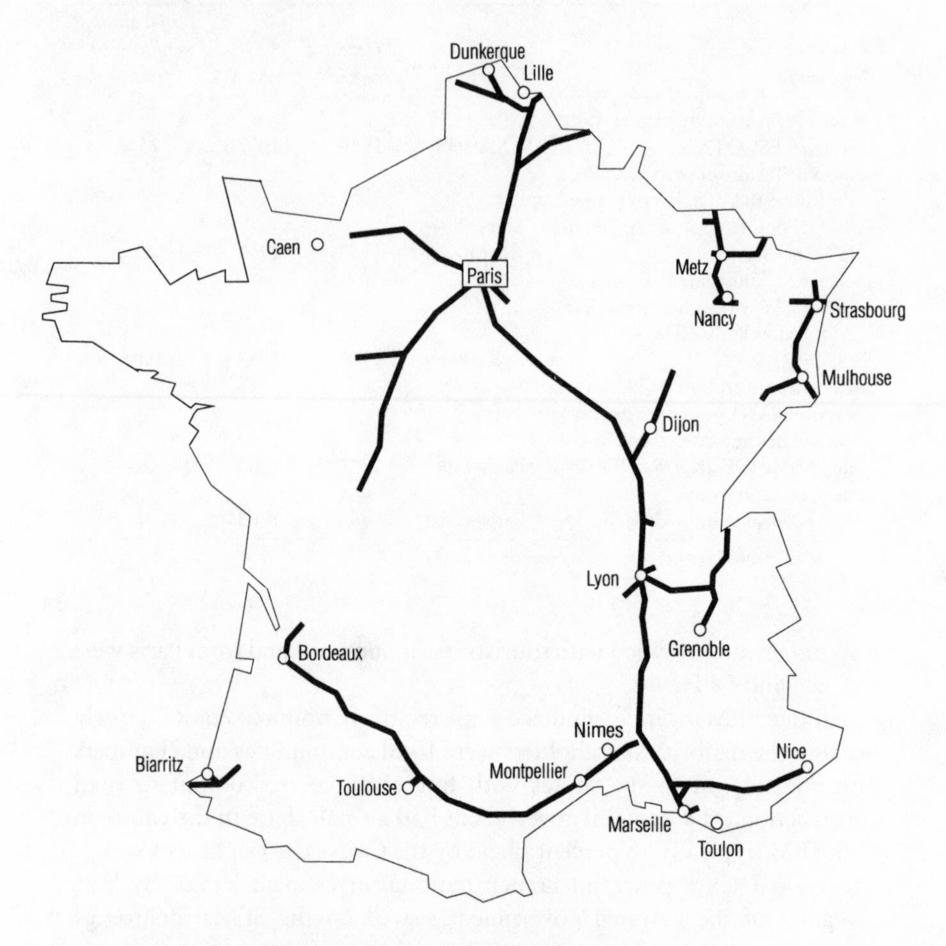

SOURCE: Highways Directorate, Ministry of Equipment, Paris, 1960.

ing losses if needed. The cash construction and balancing advances were to be repaid, but without interest and only after cash flow was sufficient to service debt, pay operating expenses, and build up adequate reserves. The construction cash advances provided for SEM concessions granted during the late 1950s and 1960s averaged 30 percent to 40 percent of construction costs.[8] No estimates are available for in-kind advances, although

8. Fayard, *Les Autoroutes*, pp. 82–87 .

these were apparently important in a few cases.[9] Balancing advances totaled 100 million francs during the 1960s, with most of the funds required by the companies building Paris-Lyon (SAPRR) and Paris-Lille (SANEF).[10]

Throughout the 1960s the SEMs were little more than paper organizations. The CDC marketed the state-guaranteed loans for the SEMs through a special office established in 1963: the Caisse National des Autoroutes (CNA). Another subsidiary of the CDC, the Société Centrale pour l'Equipement du Territorie (SCET) managed the accounts and works contracts. The Highway Directorate in the national Ministry of Equipment designed, built, and maintained the autoroutes for the SEMs, and the head of the Highway Directorate served as the chairman of the CNA and the national government's delegate to the SEM boards. The SEMs maintained a small office in Paris, but their main task was simply to staff the toll plazas, collect tolls, and pay off the debt.

The rate of construction fell short of that contemplated in the 1960 master plan. Only 1,125 kilometers of intercity autoroutes were in service by the end of the 1960s, although this included the important Marseille-Lyon-Paris-Lille spine. The primary obstacle to faster construction was the Ministry of Finance that, by law, controlled the debt that the SEMs and other public enterprises could issue. The ministry generally favored private over public sector investments and reportedly placed a low priority on public roadbuilding at the time. Highway traffic increased rapidly throughout the 1960s, however, challenging the Ministry of Finance's priorities and provoking another round of reforms.

Private Concessions and Accelerated Construction, 1969–81

President Georges Pompidou, who took office in 1969, placed a higher priority on road development than his predecessor, Charles DeGaulle. Pompidou appointed an aggressive new minister of equipment, Albin Chalandon, who soon modified the autoroute program in three important ways: allowing private companies to compete for new concessions, giving the existing SEMs more autonomy and responsibility, and issuing a new master plan that called for building 3,000 kilometers of new

9. Fayard, *Les Autoroutes*, p. 78.
10. Fayard, *Les Autoroutes*, p. 77.

autoroute in the 1970s, an almost threefold increase over the construction rate of the 1960s.[11]

Chalandon hoped that private concessions would provide new sources of financing, have lower construction and operating costs, and stimulate improved performance from the SEMs. Chalandon held four competitions for new private concessions beginning in 1969, resulting in awards to four new private toll road companies between 1970 and 1973: Cofiroute was awarded 462 kilometers of route from Paris south to Poitiers and west to Le Mans; AREA was to build autoroutes east of Lyon toward the Alps; APEL was to build an autoroute from Paris east to Metz; and ACOBA was awarded a short (sixty-three-kilometer) segment of autoroute extending north from Bayonne (near the Spanish border) toward Bordeaux (table 8-3).

All four new concessionaires were consortia of major French public works construction companies and French banks. Toll roads were thought to be a very long-term investment, with payback expected over twenty-five years. Apparently, few investors were willing to wait so long for their return unless they saw the prospect of other benefits as well. The banks generally held a comparatively small share of the stock, reportedly purchased because they wanted to earn fees issuing the bonds rather than because they wanted investments in their own right. The construction contractors, who were usually the majority shareholders, could wait for a return from their investment because they also expected to build the autoroute and to earn a reasonable profit from doing so.[12]

The Ministry of Equipment selected the route for each competition but left the competing consortia free to propose detailed designs and financial, management, and operating plans.[13] A jury of officials from the Ministries of Equipment and Finance judged the proposals based on the size of the cash construction advances requested from the state, which could not exceed 25 percent of the cost; the equity and reserves committed by shareholders, which had to be at least 10 percent; the loans not

11. This account of the Chalandon reforms is based primarily on Fayard, *Les Autoroutes*, pp. 53–65; Bernard Seligman, "General Presentation of the French Highway System," pp. 18–20, in *France*, vol. 2, *Toll Roads: Report of a Seminar-cum-Study Tour* (Bangkok: United Nations, Economic Commission for Asia and the Pacific, 1985); and interviews with Cyna, Leyrit, and Lafont cited earlier.

12. To mollify small construction companies, the terms of the competition stated that the concessionaire should be free to select a construction company of its choice; Fayard, *Les Autoroutes*, pp. 57, 60. In practice most contracts were apparently awarded to shareholders.

13. Fayard, *Les Autoroutes*, pp. 61–62.

TABLE 8-3. The Four Original French Private Concessionaires

Company	*Intended route*	*Date created*	*Initial equity (millions of francs)*	*Current status*
Compagne financière et industrielle des autoroutes (Cofiroute)	Paris–Poitiers, Paris–Le Mans	1970	462	Private
Société des autoroutes Rhône–Alpes (AREA)	Alpine autoroutes	1971	359	SEM
Société des autoroutes Paris–Est–Lorraine (APEL)	Paris–Metz	1972	315	Absorbed by SANEF
Société des autoroutes la Côte Basque (ACOBA)	Hendaye–Bayonne	1973	63	SEM owned by ASF

SOURCE: Fayard, *Les Autoroutes et Leur Financement*, pp. 25–40.

guaranteed by the state, which had to be at least 15 percent; the quality and reliability of the cost and traffic projections; and, in the first two competitions, the speed with which the autoroute would be opened.[14]

Between two and four consortia entered each competition. Consortia that had lost earlier competitions were often successful in subsequent ones, which left the impression that the government tried to offer enough concessions so that every interested consortium would eventually be awarded one. The SEMs did not compete directly, although at least once an SEM successfully offered to pick up a segment of a proffered route in which the private bidders were less interested.[15]

Chalandon also gave the SEMs more independence and autonomy in an effort to improve their performance and responsibility. He made the SEMs responsible for designing, constructing, and maintaining their roads as well as for collecting tolls. The SEMs were awarded significant new routes or segments as well, although not through a competitive process like that applied to the private concessions. Usually the new concessions were for branches off the main trunk routes, particularly if those trunk routes were beginning to carry heavy traffic and generate substantial surpluses. The SEMs often received an extension of the date at which their original concessions would expire in return for assuming a new concession.

14. Fayard, *Les Autoroutes*, pp. 54–55.

15. This occurred when the Paris-Strasbourg route was offered for competition, and private bidders were most interested in Paris-Metz only. SANEF offered to take and was awarded Metz-Strasbourg; Fayard, *Les Autoroutes*, pp. 56–57.

The government also became less generous with assistance for any new concessions, whether to SEMs or private firms, in keeping with Chalandon's desire to promote responsibility among the concessionaires and the Ministry of Finance's desire to control costs. Beginning in 1970, construction advances were made only for segments the government deemed unprofitable and new advances would have to be repaid with interest.[16] No more than 70 percent of debt could be in loans guaranteed by the state. Nevertheless, significant aid was made available in the early 1970s to both private and SEM concessions, though not quite on the previous scale.[17]

The oil price shocks of the 1970s created serious problems for many of the new concessions by driving up roadbuilding costs just as they were getting under way. The French government's construction price index increased by 20 percent a year for several years in the mid-1970s. Typical autoroute construction costs, which had hovered around 4.3 million francs per kilometer through the early 1970s, increased to 10 million francs per kilometer by 1978.[18] Interest rates went up as well, which increased the financing costs for the new projects.

These problems were compounded by a slowdown in traffic growth and the government's reluctance to raise tolls. Between 1970 and 1975 the average traffic per kilometer had increased by an amazing 71 percent on the Lille-Paris-Lyon-Marseille spine and by 39 percent for the five SEMs as a whole (rising from about 13,000 vehicles to about 17,000 vehicles per kilometer each day). Starting around 1974 average traffic density leveled off and only began to grow again in the late 1970s, and then at a much slower rate.[19]

The Ministry of Finance's desire to control the inflation caused by the energy crisis led it to hold down the rate of toll increases. The original concession agreements of the five SEMs specified that the Ministry of

16. The interest rate was a fraction of the inflation of an index of road construction costs.

17. For example, the initial plans for the four private concessions called for, on average, 12 percent of construction costs to be covered by state cash advances and 52 percent by state-guaranteed loans. Of the balance, 22 percent was to be covered by loans without guarantee, 8 percent by equity, and 6 percent by "other resources"; Fayard, *Les Autoroutes*, pp. 81, 90. The SEMs also continued to receive cash construction advances of 10 percent to 50 percent for some new segments, although some substantial segments were granted without any advances; Fayard, *Les Autoroutes*, pp. 82–87 .

18. Fayard, *Les Autoroutes*, pp. 69, 70, 72; and Seligmann, "The French Highway System," p. 20.

19. Fayard, *Les Autoroutes*, p. 107.

Finance would approve toll rates, but the agreements of the four private companies stated that private concessionaires could set their own toll rates for an initial period (in Cofiroute's case, the first ten years of the concession) and the Ministry of Finance would set them thereafter. In 1975 the Ministry of Finance declared that it would regulate private as well as SEM tolls; the private companies sued the government for breach of contract but eventually lost.[20] Autoroute tolls, once firmly under the Ministry of Finance's control, rose at only about half the rate of inflation through the 1970s.[21]

By the end of the 1970s, 5,235 kilometers of autoroute were in service including 4,054 intercity kilometers, mostly on tolled concessions. Many of the concessionaires were in serious financial difficulty, however, especially the four private ones since they had built all of their projects in the 1970s. Some of the SEMs were also in trouble too, since the benefits of having built heavily traveled roads in the 1960s were partially offset by their commitments to build more lightly traveled branches in the 1970s.

State Takeovers and Cross Subsidies, 1981–87

A new round of reforms began in 1981 when the conservative government of President Giscard-D'Estaing was replaced by the socialist government of President François Mitterrand. The new minister of equipment, Charles Fiterman, was a communist in Mitterand's coalition and reportedly wanted to abolish tolls on the autoroutes altogether. He soon realized that abolishing tolls was financially impractical and decided instead to take over three of the four ailing private concessionaires, to reduce differences in toll rates among the SEMs, and to implement a formal system of cross subsidies among the SEMs.[22]

Fiterman had been inclined to take over all four private concessions, but a special report he commissioned convinced him that Cofiroute was financially viable, well managed, and worth preserving. Of the remaining

20. A French court ruled that the concession agreements violated a 1945 law that declared the government had the power to regulate any price; Cyna, interviews.

21. Tolls were increased at about half the rate of general price inflation and only about one-third the rate of construction price inflation; Fayard, *Les Autoroutes*, p. 112.

22. This account of the reforms of the early 1980s is based primarily on Seligmann, "The French Highway System," pp. 20-22; Gerard Morançay, "Le Financement des Autoroutes Françaises," *Les Cahiers de l'Administration: Special Autoroute* (Paris: Administration, 1990), pp. 19–31; and the interviews with Cyna, LaFont, Tennenbaum, Lecomte, and Leyrit cited earlier.

three private concessionaires, one was absorbed by a neighboring SEM (APEL, which was taken over by SANEF), while the other two were converted to new SEMS (ACOBA and AREA). The result was the present system of seven autoroute SEMs (the five originals plus ACOBA and AREA), one private autoroute concession (Cofiroute), and two special-tunnel SEMs.

Observers offer several reasons why Cofiroute survived while the other three privates failed. Cofiroute's concession is regarded as the best of the four, with more reasonable construction costs and higher traffic potential.[23] Cofiroute benefited from being the first private concession as well, so it had a few more years to build before the energy crisis hit.

Perhaps equally or more important, Cofiroute is widely regarded as well managed and was reputedly more resistant than others to potential conflicts of interests with its construction company shareholders. For example, Cofiroute came in on budget with its construction costs while the other three did not, although this success was surely partly because of Cofiroute's earlier start. Given the modest equity contributions required and the high proportion of advances and state-guaranteed loans, some observers have suggested that the principal shareholders of the other three private companies were more interested in construction contracts than in operating an autoroute. This suspicion is furthered by their acceptance of apparently less lucrative concessions, rapid construction, and cost overruns.

The second element of Fiterman's reforms was a policy of toll harmonization. The initial toll rates had been fixed at the time each concession was granted, and concessions for more costly segments were usually granted a higher toll per kilometer traveled. Under Fiterman's reforms, tolls were not made completely equal, in recognition that difficult terrain might justify higher tolls, but the differential between the most and least expensive autoroutes was reduced from 3 to 1 to 2 to 1.

The final and most critical reform was the creation in 1982 of a new government agency, Autoroutes de France (ADF), to serve as a clearing-house for issuing new advances to and receiving payment of old advances

23. APEL's Paris-Metz concession does not take a direct route west from Paris to Nancy or the German border, for example, but was routed north by the Ministry of Equipment to serve Reims and Metz for political reasons; as a result, APEL lost some of its potential traffic to a parallel national route (N-4) which, though not built to autoroute standards, is more direct. ACOBA's route at Bayonne had modest traffic potential and was offered as a concession largely for political reasons, while APEL's alpine routes suffer from high construction costs.

from the SEMs. All SEM debts for advances from the state were transferred to ADF, and the funds received from the SEMs were used to pay advances to those SEMs still operating at a loss. The government had been pursuing a policy of cross subsidies within companies since at least the mid-1970s by giving new and less profitable sections to the companies with older and more heavily traveled segments. With ADF's creation, the government could engage in cross subsidies among companies as well, a practice made necessary by both the government takeover of the three failing private companies and the new policy of toll harmonization (since in practice toll harmonization increased the profits of the older SEMs and the deficits of the later ones). As shown in table 8-4, the ADF clearinghouse as a whole began to show growing profits during the late 1980s, helped in large part by the resumption of traffic growth and a slowdown in the granting of new concessions in the early and mid-1980s.

The reforms of the early 1980s also formalized a gradual shift in government policy. The 1960 master plan emphasized the traffic and profitability of individual segments as the appropriate criterion for deciding what autoroutes ought to be built, and thereby committed the government to toll financing of autoroutes. Chalandon's reforms of 1969–70 reinforced the reliance on tolls by calling for concessions to private companies, increasing the independence and accountability of the SEMs, and reducing advances and guaranteed loans. But at the same time, and probably unknowingly, his decision to triple the rate of new construction gradually led to a system of cross-subsidizing segments within companies and undermined the focus on the profitability of the individual autoroute segment. This process was hastened by the energy crisis but probably would have occurred inevitably as the companies expanded their systems to ever more lightly traveled branches. Of course, the reforms of the 1980s extended this development even further by establishing a system of cross subsidies among the SEMs as well as within them.

Continued Expansion with Second Thoughts, 1988–92

The policies established in the early 1980s were reaffirmed in 1988 and 1990 with new plans for dramatic expansions of the intercity autoroute network. The 1990 plan called for an eventual intercity system of 9,530 kilometers (figure 8-2), with construction to proceed at 350 kilometers a year, triple the rate of the early 1980s and double that of the

TABLE 8-4. Advances and Repayments between Autoroutes de France and Sociétés d'Économie Mixte, 1984–90
Millions of francs

Year	*SEM surpluses*	*SEM deficits*	*Net*	*Repayment of ADF to state*	*Debts of ADF to state*	*Debts of SEM to ADF*
1984	444	316	128	...	6,217	6,088
1985	547	542	5	...	8,134	8,002
1986	345	805	(460)	...	10,375	10,691
1987	215	356	(141)	...	10,717	11,192
1988	685	216	469	...	10,895	10,925
1989	1,079	171	908	660	11,144	10,302
1990	1,499	67	1,432	1,432	9,466	9,003

SOURCE: Autoroutes de France, *Annexe Compatable*, appendix to *Rapport du Conseil d'Administration sur l'Activité 1990* (Paris: 1991), pp. 11–18. SEM debts to the state increase annually in part because of interest charged on post-1970 advances. The reported reduction in ADF debts to the state between 1989 and 1990 seems inconsistent with the reported repayments of ADF to the state during 1990.

mid-1980s.[24] As before, cross subsidies were to be a principal source of financing new segments. Indeed, the rate of construction in the new plan was apparently determined by a projection of the cross subsidies available within and among the SEMs assuming reasonable traffic growth, tolls that kept pace with inflation, and an extension of the concessions to the year 2015.[25]

Serious threats to accelerated construction have been posed, however, by the Ministry of Finance's desires to capture the ADF clearinghouse's surpluses for nonhighway purposes and to limit the value of toll increases. In 1989 and 1990, for example, the Ministry of Finance forced the ADF to transfer 2.1 billion of the 2.4 billion francs in surpluses it earned to reduce its debts to the state, leaving only 0.3 billion francs in new surpluses for road construction. The Ministry of Finance has also been reluctant to grant toll increases, because it is more concerned with the problems of inflation and monopoly than with the rate of autoroute construction. Uncertainties about future toll increases have been a serious obstacle to new concessions to private autoroute companies. In the early 1990s, for example, the Ministry of Equipment pressed Cofiroute to accept a new intercity concession, but Cofiroute refused unless the Ministry of Finance provided assurances about future rates of toll increase.[26] The SEMs have been willing to accept new segments without future toll

24. Autoroutes de France, *Rapport du Counsel D'Administration sur l'Activité 1990)* (Paris: Autoroutes de France, 1991), p. 7.

25. Morançay, *Le Financement*, p. 21.

26. Cyna and Leyrit, interviews.

FIGURE 8-2 France's 1990 Autoroute Masterplan

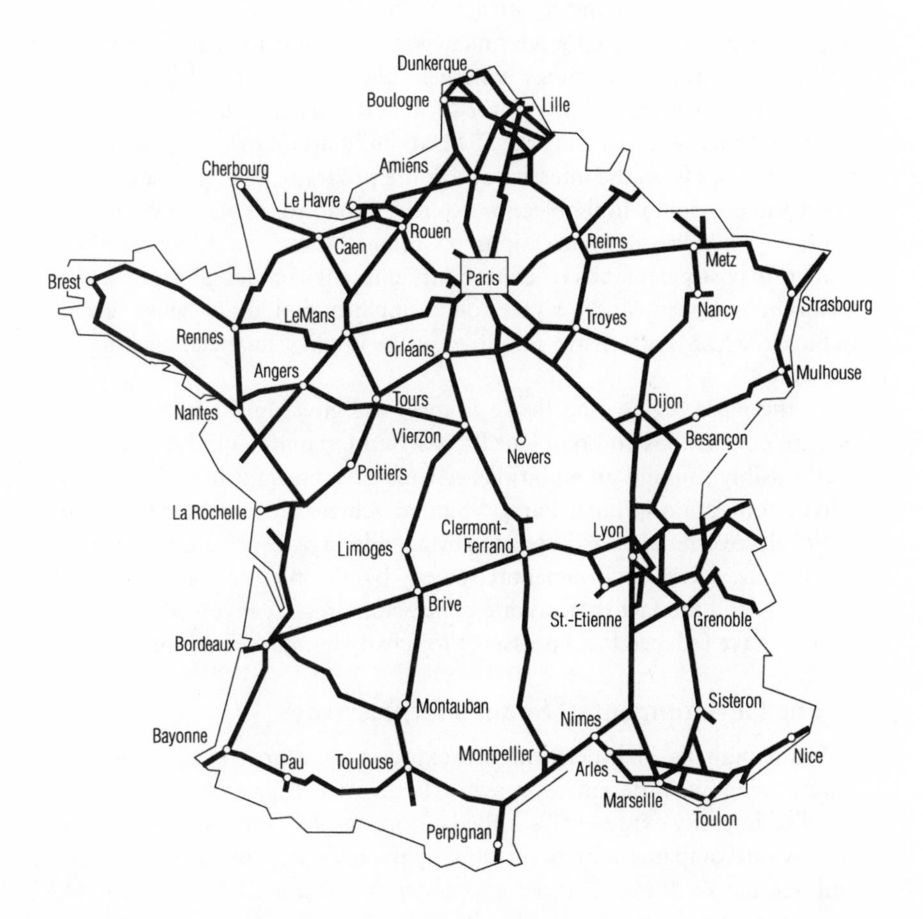

SOURCE: Highways Directorate, Ministry of Equipment, Paris, 1991.

guarantees, but they will not be able to build at planned rates if the Ministry of Finance grants toll increases at less than the level of inflation.

While the intercity autoroute program continued to rely primarily on the SEMs, national and local governments began to consider the use of private concessions for urban expressways. Urban expressways had not been tolled in the past, largely because of fear that tolling would be politically difficult on routes used for everyday commuting. Beginning in the late 1980s, however, new private urban toll roads were being considered in the Paris, Lyon, and Marseille metropolitan areas. The high cost of building certain badly needed autoroute segments in major metropolitan areas made toll financing attractive. Private concessions were attractive because the national government was reluctant to require that the SEMs build urban autoroutes, whether tolled or untolled. Urban autoroutes were the responsibility of the local and departmental governments in urban areas, and requiring the SEMs to build urban roads would divert their surpluses from the intercity autoroute program. Growing neighborhood and environmental objections to new urban roads were a consideration as well. Private concessionaires might be more able to resist or moderate costly demands of community and environmental groups, the government hoped, since private concessionaires would not be perceived to have the SEMs' extensive resources or the implicit financial backing of the state.[27]

In the Paris region (the Ile de France) the government hopes to use private concessions to complete Paris's two circumferential autoroutes and possibly to build an ambitious scheme of subterranean toll roads to relieve congestion in inner Paris.[28] Similar schemes are being negotiated by local governments in Lyon and Marseille. These private urban toll roads must clear environmental review by the national government, however, and the fact that private concessionaires are involved does not seem to have reduced local pressures for costly mitigation measures.

The Development of Spain's Expressways

Spain built its high-performance expressway system in two distinct stages, using two different approaches. In the 1960s Spain began a system of tolled intercity expressways, called autopistas, by granting concessions to private companies. In the 1980s Spain reversed course and began building untolled, tax-financed expressways called autovías rather than expand the autopista concessions. By 1990 approximately 2,000 kilometers of each type had been opened (see table 8-5). The toll autopistas built during the late 1960s and 1970s are concentrated around Spain's major centers of economic activity and traffic (primarily in the northeast, between Barcelona and Bilbao, and along the Mediterranean coast), while the untolled autovías built during the 1980s fill in the national network (primarily with roads radiating out of Madrid; see figure 8-3).

27. Leyrit, interview.

28. "Les Rocades Autoroutiers des Grandes Métropoles; Le Cas de Paris et Lyon," *Bulletin des Autoroutes Françaises*, no. 33 (Association des Sociétés Françaises d'Autoroutes, June 1991); and Peter Reina, "French Mega-builder Bouygues Targets Concessions at Home," *Public Works Financing*, May 1991, pp. 20–21.

TABLE 8-5. Kilometers of Autopista and Autovía in Service, Selected Years, 1960–90

	Concessionary toll autopistas					
Year	*Total conceded*	*Under development*	*In service*	*Other autopistas in service*	*Autovías in service*	*Autopistas and autovías in service*
1960	0	0	0	0	0	0
1967	167	167	0	n.a.	n.a.	n.a.
1970	508	426	82	n.a.	n.a.	n.a.
1975	2,042	1,426	616	n.a.	n.a.	n.a.
1980	2,042	420	1,622	37	273	1,933
1985	2,042	207	1,835	299	959	3,093
1989	2,113	236	1,877	309	2,006	4,292
1990	2,148	253	1,895	n.a.	n.a.	n.a.

SOURCES: Concession figures for 1967–90 from Ministerio de Obras Públicas y Transportes, Delegación del Gobierno en las Sociedades Concesionarias de Autopistas Nacionales de Peaje, *Memoria 1990* (Madrid: Ministerio de Obras Públicas y Transportes, 1991), p. 11. Figures for all autopistas and for autovías for 1980–89 from Ministerio de Transportes, Turismo y Communicaciones, *Informe Annual: Los Transportes, el Turismo y las Comunicaciones 1989* (Madrid: Ministerio de Transportes, Turismo y Comunicaciones, 1990), p. 64. The figures for "concessionary toll autopistas" reported in the first source (and reproduced above) are twenty to forty kilometers greater than figures reported for "toll autopistas" in the second source; the discrepancy may be because of the presence of short untolled segments in the concessions.
n.a. Not available.

In part because of its mixed approach, Spain has been more successful than France in maintaining a private toll road industry. The autopistas are operated by thirteen companies; twelve began as private concessions. Nine of the twelve still are private, while three were taken over by the national or regional governments in the 1980s. The two largest private companies dominate the entire autopista industry, together carrying 70 percent of all the traffic on toll autopistas and operating 52 percent of the route kilometers. By contrast, the public autopista concessionaires together account for only 7 percent of traffic and 16 percent of the route kilometers.[29]

Toll Autopistas Built by Private Concessionaires, 1960–81

Toll highways had been authorized as early as 1953 in Spain, and in 1960 the Spanish government let its first private concession for a short

29. Only three of the four public companies were operating autopistas in 1990; the fourth (TABASA) was still building its newly granted concession. Ministerio de Obras Públicas y Transportes, Delegación del Gobierno en las Sociedades Concesionarias de Autopistas Nacionales de Peaje, *Memoria 1990* (Madrid: Ministerio de Obras Públicas y Transportes, 1991), pp. 7, 25.

FIGURE 8-3 Spanish Autopistas in Service, 1991

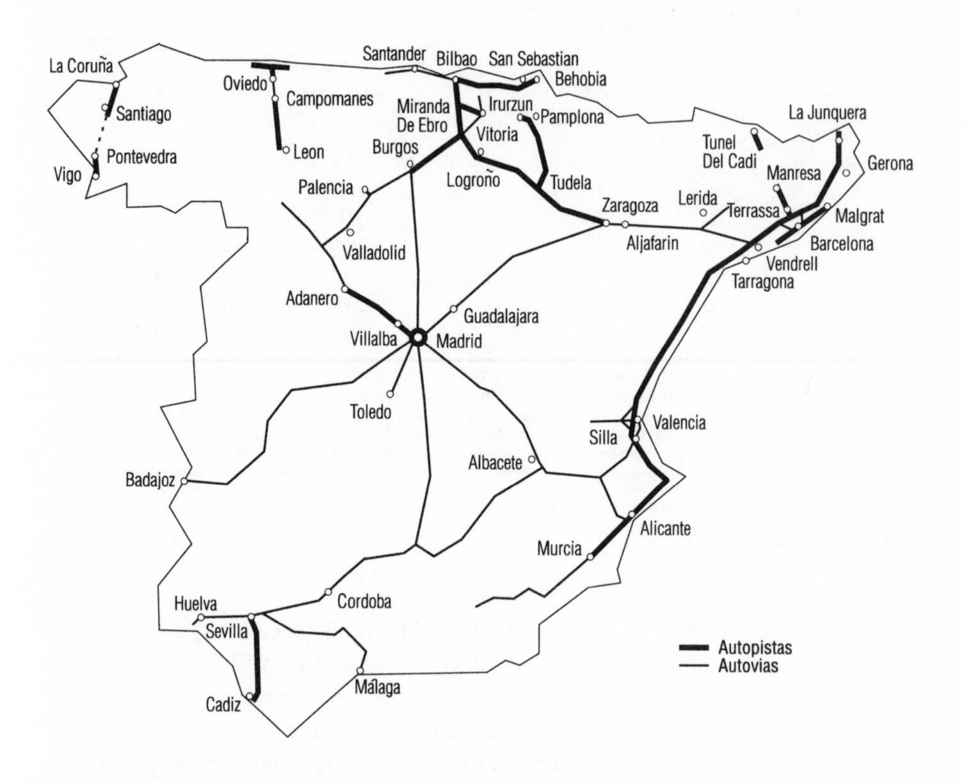

SOURCE: Asociación de Sociedades Españolas Concessionarias de Autopistas, Túneles, Puentes y Vias de Peaje, "Las Autopistas de Peaje de España," 1991.

toll tunnel through the Guadarrama mountains to the northwest of Madrid.[30] Spain's government did not begin to seriously consider the

30. This account of the origins of the autopistas is based primarily on José Luis Ceron-Ayuso, chairman and managing director, Autopistas del Mare Nostrum (AUMAR), interview, September 30, 1991; José Maria Jerez, director, Asociación de Sociedades Españolas Concesionarias de Autopistas, Túneles, Puentes y Vías de Peaje (ASETA), personal interview, September 30, 1991; ASETA, "Las Autopistas de Peaje en España," 1991; José Luis Ceron-Ayuso, "The Toll Highways in Spain and an Approach to the Future," pp. 83–88, in *Presentations* (Washington: International Bridge, Tunnel, and Turnpike Association, 1981); and Organization for Economic Cooperation and Development, *Toll Financing and Private Sector Involvement in Road Infrastructure Development* (Paris: OECD, 1987) pp. 98–101. For a detailed history of the Spanish expressway experience in English see Gómez-Ibáñez and Meyer, *Private Toll Roads in France and Spain*, pp. 49–76.

construction of an extensive system of high-performance toll expressways until the early 1960s , however. Growing congestion on Spain's network of two-lane national roads finally led, in 1967, to a national highway plan that called for the construction of 4,800 kilometers of expressways, or autopistas, by 1985.[31] Near and in major cities, the autopistas would be built by the national government and not tolled. Intercity autopistas would be built by private concessionaires and financed with tolls.

The decision to use tolls and private concessionaires on the intercity autopistas was apparently not controversial at the time. Tolls seemed an obvious choice. Spain's tax resources were limited, and there were many competing uses for public funds. France and Italy served as examples, since they had already embarked on their programs of intercity toll road construction. Furthermore, only Spain's already developing or industrialized areas—mainly around the cities of Barcelona, Madrid, Sevilla, and Bilbao—would have the traffic volumes necessary to justify high-performance roads in the near future; it seemed fairer to ask those wealthier regions to pay for their own expressways through tolls than to finance the expressways from national tax revenues. Finally, foreign tourists were expected to be heavy users of some of the expressways, especially along the Mediterranean coast, and tolls would help ensure that they paid their share of Spain's infrastructure costs.

Concessions were granted to five private companies between 1967 and 1972. Two private companies (ACESA and AUMAR) were awarded concessions to build a road from the French border to Barcelona and down the Mediterranean coast to Alicante. The private company that had been awarded the Guadarrama tunnel concession in 1960 (Iberpistas) was allowed to build seventy kilometers of autopista on either side of its tunnel. Two other new private companies (Europistas and Biteca de Autopista) were given concessions on the approaches to two of Spain's other major cities (Bilbao and Sevilla).

This early experience was so encouraging that in 1972 the government formalized and standardized concession terms under a new law of autopista concessions.[32] Many of the key provisions dealt with concession financing, particularly requirements that foreign debt be used where

31. Ceron, "Toll Highways in Spain," p. 84.

32. "Ley de construcción, conservación y explotación de autopistas en régimen de concesión," Ley 8/1972, de 10 de mayo, *Buletin Oficial del Estado*, no. 113 (May 11, 1972), pp. 8229–34; and "Pliego de clausuales generales para la construcción, conservación y explotación de autopistas en regimen de concessión," *Buletin Oficial del Estado*, no. 41 (February 16, 1973), pp. 2998–3011.

possible. The Spanish government had required the early concessions to finance a large part of their costs from foreign debt in order to ease Spain's balance-of-payments problems and to avoid drawing away domestic savings from other projects. The 1972 law set standards that at least 45 percent of construction costs be financed from foreign loans, at least 10 percent from equity, and no more than 45 percent from domestic loans. The early Spanish autopista companies had trouble raising funds from foreign capital markets, however, and in return the government agreed to guarantee some of these loans and to protect the companies from exchange rate fluctuations. The 1972 law specified that the government would guarantee up to 75 percent of the foreign loans; moreover, all foreign loans would be denominated in pesetas with the government assuming the full exchange rate risk.[33] Initial toll rates were specific to each concession, but future toll increases would be governed by a formula depending equally on the rates of price inflation for three factors: steel, petroleum, and labor.[34] An office of the Delegate of the Government for Autopistas was established in the Ministry of Public Works to monitor the compliance of the concessionaires with the 1972 law and its accompanying regulations.

The government also accelerated the autopista construction program, issuing in 1973 a new plan calling for an eventual system of 6,594 kilometers of autopistas, an increase of more than 2,000 kilometers over the 1967 target.[35] Major new concessions to seven new private toll road companies were granted between 1973 and 1975.[36] Two companies (AVASA and ACASA) were to build a 500-kilometer toll road across the northeastern half of Spain linking the industrial city of Bilbao with the

33. The 1972 law also established other special financial conditions. A special reversion fund was also required to insure that when the autopista reverted to the government at the end of the concession period, the company would have sufficient funds to pay off all debt and equity. Essentially this reversion fund would function like a depreciation reserve. To insure that the companies were always well capitalized, the companies are also required to maintain special reserves for equity investors. The two reserves give stock in Spanish toll road companies some of the characteristics of bonds, since the stockholders receive dividends during the life of the concession plus a cash distribution at the end of the concession. See Gómez-Ibáñez and Meyer, *Toll Roads in France and Spain*, p. 45.

34. The original February 1973 regulations implementing the 1972 law do not appear to give equal weight to these three price indices. By 1981, however, the formula had apparently been revised to give them equal weight.

35. The plan was issued in 1973; see Ceron, "Toll Highways in Spain," p. 85; and Ceron and Jerez, interviews.

36. Two other short-lived concessions were also granted in this period.

Mediterranean coast. The remaining five companies were scattered across northern Spain.[37] With these concessions, a total of twelve private companies had received awards amounting collectively to around 2,000 kilometers (table 8-6). These concessions would account for almost all of the autopista routes in operation by 1991.

It is unclear whether the concessions granted in the late 1960s and early 1970s were awarded competitively or not. In theory, the government would consider alternative proposals and select the concessionaire on several criteria, including the equity invested, the toll rate proposed, the duration of the concession (to a maximum of fifty years), and the guarantees or subsidies requested.[38] Apparently often only one company was seriously interested, and the competition took the form of a negotiation between that company and the government.

All but one of the companies had taken advantage of the government guarantees for a portion of their foreign loans and of government coverage of the exchange risk. Almost no other subsidies were provided, however, even though the autopista law allowed the government to grant subsidies or advances if it deemed them necessary.[39]

The death of General Francisco Franco in 1975 and the energy crises during that decade brought a fundamental reconsideration of Spain's autopista program. The government's plans to award additional concessions were postponed with Franco's death. In any event, the energy crisis soon caused serious financial problems for many of the private concessionaires, making further awards more difficult. As in France, those concessionaires that had most recently begun construction were most affected by the resulting cost inflation and traffic downturn.

37. All but the last of these companies received concessions from the national government. AUDENASA was the first and for about fifteen years the only concession granted by one of Spain's regional governments: the government of Navarra.

38. Carlos Martin-Plasencia, delegado del Gobierno en las Sociedades Concesionarias de Autopistas Naciónales de Peaje, Ministerio de Obras Públicas y Transportes, interview, Madrid, December 5, 1991; Francisco Fernandez-Lafuente, director general de Planificación Intermodal del Transporte en las Grandes Ciudades, Ministerio de Obras Públicas y Transportes, interview, Madrid, December 5, 1991; and Ceron and Jerez, interviews.

39. Europistas did receive a government guarantee of traffic volumes at some point, however. This guarantee called for the government to pay if traffic volumes fell short of 20,800 vehicles a day, although the government's obligation was limited to the toll revenue from 2,500 vehicles a day. The guarantee ended in 1991; Martin, interview.

TABLE 8-6. Spanish Autopista Concessions, 1967–75

Company	*Date*	*Concession*	*Kilometers*
Autopistas Concesionaria Española (ACESA)	1967	Barcelona–French frontier	150
	1967	Mongat–Mataro	17
	1968	Barcelona–Tarragona	100
	1974	Montmeló–El Papiol	27
Ibéria de Autopistas (Iberpistas)	1968	Madrid northwest to Guadarrama tunnel concession of 1960	70
Europistas	1968	Bilboa–French frontier	106
Biteca de Autopista	1969	Sevilla–Cádiz	94
Autopistas del Mare Nostrum (AUMAR)	1971	Tarragona–Valencia	225
	1972	Valencia-Alicante	148
Autopistas de Cataluña y Aragón (ACASA)	1973	Zaragoza–Mediterráneo	216
Autopista Vasco-Aragonesa (AVASA)	1973	Bilboa–Zaragoza	294
Sociedad de Autopistas de Navarra (AUDENASA)	1973	Tudela–Iruzun	113
Autopistas del Atlántico (AUDASA)	1973	Ferrol–Portugese frontier	213
Túnel del Cadí	1973	Tunnel near France	30
Eurovías	1974	Burgos–Cantábrico	150
Astur–Leonesa (AUCALSA)	1975	Campomanes–León	87

SOURCE: Asociación de Sociedades Españolas Concesionarias de Autopistas, Túneles, Puentes y Vías de Peaje, "Las Autopistas de Peaje de Espana," Madrid, 1991.

Untolled Autovías and Consolidation of the Autopistas, 1982–87

The election of the Socialist government of Felipe Gonzales in 1982 brought a rethinking of Spanish highway policy. A policy review, completed in 1984 and presented in a national highway plan for 1984–91, argued for a significant expansion of the nation's high-performance highway network through the construction of untolled autovías rather than further expansion of the tolled autopista concessions.

Autovías versus autopistas. The shift from autopistas to autovías was reportedly a political as much as a technical decision. Autopistas were suspect to the new socialist government because they were a policy initiative of the rightist Franco regime. Tolling also seemed less egalitarian, since it might limit the use of the roads to the well-to-do.

The national highway plan dwelt only on the technical advantages of autovías over autopistas, however. In the first place, the Ministry of

Public Works argued, the roads paralleling the autopistas were still heavily traveled while the autopistas were underutilized. This situation occurred not only because autopistas were tolled but also because they were not designed for, and thus unattractive to, motorists traveling short and medium distances.[40]

More important, autovias could be built at lower cost than autopistas, the ministry argued. The key to the savings was that autovías could use the existing right-of-way and roadway of the present national roads. Autopistas, by contrast, could be built only if a free parallel route was available, and thus required an entirely new right-of-way. Where possible, the ministry planned to use existing two-lane national roads for one direction of traffic and to build a second, separate two-lane carriageway alongside for traffic in the opposite direction. This dual-carriageway design would mean some compromises with typical autopista standards, such as tighter curves, shorter ramps, and occasional left-lane exits or at-grade crossings. Costs per kilometer would be reduced by one-third, however, according to ministry estimates.[41] The lower cost design would, in turn, allow the government to build a more extensive national network than it could otherwise and thereby "reduce the regional imbalances in investments in high-capacity highways."[42]

The highway plan for 1984–91 identified 5,700 kilometers of routes that carried sufficient traffic to warrant an autopista or an autovía. Three hundred thirty billion pesetas (approximately U.S. $3 billion at exchange rates of the time) would be spent to build 3,250 kilometers of new autovía to supplement the nearly 2,000 kilometers of existing autopista concessions.[43]

The autopista industry initially fought the government's decision to build autovías. The dual-carriageway design would prove inadequate as traffic volumes grew, the industry argued, and eventually a wholly new route would have to be built at double the cost. The autopista companies soon realized that they had much to gain from the program, however, since the new autovía network was designed for the most part to comple-

40. Ministerio de Obras Públicas y Urbanismo, *Memoria Resumen Plan General de Carreteras 1984–91* (Madrid: Ministerio de Obras Públicas y Urbanismo, 1984), p. 12.

41. "The Motorway Industry in Spain," p. 5.

42. Ministerio de Obras Públicas y Urbanismo, *Memoria Resumen*, p. 15.

43. Apparently approximately 250 kilometers of autovía were to be built by regional governments as well, primarily a road in southern Spain connecting Sevilla with Granada and Baza; see Ministerio de Obras Públicas y Urbanismo, *Memoria Resumen*, pp. 30–31 .

ment rather than compete with the autopista concessions and thus would feed traffic onto the autopistas.[44]

Autopista reforms. The government's enthusiasm for autovías was based in part on its perception that the autopista concession system was seriously flawed. Several of the later concessions were close to bankruptcy, which would require the government to make good its guarantees on their foreign loans. Even if bankruptcy could be avoided, the government's assumption of the exchange risk on foreign debt was costing about 40 billion pesetas a year, largely because the value of the peseta had fallen dramatically during the severe Spanish recession of the early 1980s.[45]

In explaining the situation to the Spanish legislature in 1984, the then-minister of public works, Julian Campo, argued that several features of the autopista concession system had served to transfer a disproportionate risk to the state.[46] In the first place, the minister contended that it had been a mistake to allow the companies to have as little as 10 percent equity investment. The undercapitalization of the autopistas meant that their annual debt service payments were nearly as much as their total shareholder equity. When the companies got into trouble, they had few resources to draw on.

Equally important, according to Campo, were the incentives created by the availability of government loan guarantees for up to 75 percent of foreign debt and government assumption of the exchange rate risk on all foreign debt. It was not surprising that 75 percent of the 459 billion pesetas in capital raised by the autopista industry by the end of 1983 had been financed with foreign debt (52 percent government guaranteed), while only 12 percent had been raised in equity and 13 percent in domestic debt.[47] The government assumption of exchange risks was pernicious, since it encouraged the companies to search for foreign loans with the lowest nominal interest rates, regardless of the exchange risks.

44. There were some exceptions, however. Autovías were proposed to parallel the ACASA's Lerida-Barcelona route (which ACESA would operate after it absorbed ACASA in 1984) and AUMAR's Alicánte-Valencia route, although the latter may not be built. Biteca de Autopista's Cadíz-Sevilla route (which AUMAR would operate after it absorbed Biteca de Autopista in 1986) would benefit from a proposed Madrid-Sevilla autovía, however. See Barclays de Zoete Wedd, "Background to the Sector," Madrid, 1990, p. 23; and "The Motorway Industry in Spain," p. 5.

45. Campo, *Las Autopistas de Peaje*, p. 9.

46. Campo, *Las Autopistas de Peaje*.

47. Campo, *Las Autopistas de Peaje*, pp. 8–9.

These difficulties were compounded by the failure to guard against the potential conflict of interest inherent when major construction companies backed toll road concessions. Campo argued it had been a mistake to allow construction companies, or banks that held a large stake in these construction companies, to be the major shareholders in the concessions, or not to prohibit the awarding of construction contracts to important shareholders. This conflict of interest contributed to overly optimistic projections of construction costs and traffic volumes when some of the concessions were awarded, Campo claimed. To be sure, he acknowledged, the energy crisis played an important role by stimulating inflation and discouraging traffic. But the discrepancies between the projections and actual results seemed too large in some cases to be explained by the energy crisis alone. In several cases, construction costs had been four or five times original projections while initial traffic volumes were as little as one-third of those expected.[48]

The old regime had abetted this excessive optimism, Campo contended, by not doing any serious independent evaluation of the traffic and financial projections offered by the companies at the time the concessions were awarded. As the companies got into trouble, moreover, the government had modified concession terms by raising toll rates, extending the period allowed for construction and the duration of the concession, and raising the amounts of external debt allowed and guaranteed.[49]

The government took over the three autopista companies in the most serious financial difficulty (AUDASA, AUCALSA, and AUDENASA) in 1984. The government subsequently paid their operating losses and granted interest-free loans to help finance the cost of completing their partly built projects.[50] Two other weak companies, ACASA and Biteca de Autopista, were merged with stronger companies, ACESA and AUMAR, in return for toll increases and extensions of the concession duration on the stronger routes. The government attempted to reduce its exchange risk by encouraging the refinancing or renegotiation of the worst of the existing foreign loans and more carefully reviewing the risks in any proposed new loans. The costs of the government's annual exchange payments declined from a peak of 83 billion pesetas in 1985 to 9 billion pesetas in 1990, although the strengthening of the peseta against other currencies in the late 1980s was probably a stronger influence than the reforms (table 8-7).

48. Campo, *Las Autopistas de Peaje*, pp. 12–13, 16–17 .

49. Campo, *Las Autopistas de Peaje*, p. 15.

50. Martin, interview.

TABLE 8-7. Government Payments for Exchange Rate Differences on Foreign Autopista Debt, 1969–90
Millions of pesetas

Year	*Annual payment*	*Cumulative payment*
1969	141	141
1970	51	193
1971	100	292
1972	70	363
1973	1,276	1,639
1974	–267	1,372
1975	64	1,436
1976	2,777	4,213
1977	5,644	9,858
1978	9,713	9,570
1979	11,033	30,604
1980	9,021	39,625
1981	17,767	57,392
1982	22,935	80,327
1983	42,382	122,710
1984	54,331	177,041
1985	83,468	260,509
1986	36,106	296,616
1987	14,944	311,560
1988	13,843	325,403
1989	7,553	332,956
1990	9,070	342,026

SOURCE: Ministerio de Obras Públicas y Transportes, Delegación del Gobierno en las Sociedades Concesionarias de Autopistas Nacionales de Peaje, *Memoria 1990* (Madrid: Ministerio de Obras Públicas y Transportes, 1991), p. 80.

The socialist government's criticism of the autopista concession system was telling but not entirely fair. Larger equity requirements might have been impossible given the pioneering and speculative nature of autopista investments in the 1960s and early 1970s . The previous government had insisted on using foreign debt for its own reasons and, at least according to the autopista companies, this was impractical at the time without some guarantees and unfair unless the government absorbed the exchange risk. The energy crises had clearly played a role in the cost overruns and the traffic shortfalls, as Minister Campo acknowledged. Nevertheless, the combination of low equity and government guarantees had clearly reduced shareholder risk and incentives to be prudent and contributed to the financial problems of the early 1980s.

Rethinking the Roles of Autopistas and Autovías, 1987–92

Interest in new autopista concessions reemerged even as the national government was building its new autovía system. The revival was initiated in the late 1980s by the regional government of Catalonia, which was concerned about rapid traffic growth and urban development in the Barcelona area as well as the approaching 1992 Olympics. Toll financing was attractive because the Catalan government budget was strained by Olympic preparations. Catalan motorists were accustomed to toll roads, moreover, since Barcelona is served by some of Spain's earliest and most successful autopista concessions (primarily operated by ACESA). The national government had programmed relatively little autovía construction in the area because of the extensive existing autopista system.

New autopista concessions near Barcelona. Between 1987 and 1990 three concessions were granted by the Catalan government and one by the national government in the Barcelona area (table 8-8). Two of the concessions granted by the Catalan government were to build a tunnel through a mountain that separates Barcelona from its growing western suburbs and to extend an autopista west from that tunnel. Because of its high cost, the tunnel was awarded to a new company owned by the Catalan regional and Barcelona city governments (TABASA). The autopista to the west was deemed sufficiently profitable to attract private interest, however, and was awarded to a new private company (AUTEMA). The remaining two concessions, for short roads along the coast north and south of Barcelona, were both awarded to private companies, one new (AUCAT) and one existing (ACESA). ACESA considered its new route so unprofitable that it insisted that the national government extend the duration of its other concessions in return.

The national government. Even as Catalonia was experimenting with new autopista concessions, the national government expanded its autovía program in several important ways. Autovía construction had fallen behind the targets set out in the 1984 highway plan and was accelerated to prepare for the 1992 Barcelona Olympics and Sevilla World Fair. A new national highway plan, adopted in 1988, increased the proposed autopista and autovía network to a little less than 6,100 kilometers and updated the cost estimates. The plan called for building 3,550 kilometers of new autovía at a cost of 641 billion pesetas between 1984 and 1992

TABLE 8-8. Autopista Concessions near Barcelona, 1987–90

Company	Date	Concession	Kilometers
New companies			
Autopista Terrassa Manresa (AUTEMA)	1987	Manresa–Terrassa	33
	1989	Terrassa–Rubí	7
Túnels i Accessos de Barcelona (TABASA)	1987	Túnel de Vallvidrera and access (Barcelona)	12
Autopistas de Cataluña (AUCAT)	1989	Castelledfels–Sitges	15
Existing companies			
ACESA	1990	Mataro–Malgrat	34

SOURCE: Asociación de Sociedades Españolas Concessionarias de Autopistas, Túneles, Puentes y Vías de Peaje, "Las Autopistas de Peaje de España," 1991.

instead of 3,250 kilometers for 330 billion pesetas between 1984 and 1991.[51]

More significantly, autovías were now to be built to standards that were essentially identical to those set for autopistas. The design compromises envisioned in the original dual-carriageway scheme were now perceived to create unacceptable safety problems as well as to reduce capacity (much as the autopista industry had predicted). The revised standards would apply not only to new autovías; some of the early autovías would be improved to eliminate the most serious problems. These design changes, together with inflation, accounted for much of the increase in projected costs.

By the early 1990s, the autovía program was on schedule but was creating a severe strain on the national budget. Of the projected 3,550-kilometer increase, 95 percent was already opened (47 percent), under construction (39 percent), or out for bid (5 percent) by June 1991. Annual government expenditures for autovía construction expenditures had exceeded 100 billion pesetas (U.S. $1 billion) every year after 1988.[52]

The high cost of the autovía program encouraged a reassessment of the government's attitude toward toll roads. After many years of negotiation, the government in 1990 agreed to revise its formula for future toll increases. The autopista industry had argued that the old formula was less relevant now that most of the concessions had been built, since

51. Ministerio de Obras Públicas y Transportes, "Informe Sobre el Primer Plan General de Carreteras," Madrid, June 1991.

52. During 1991, for example, the government expected to award 137 billion pesetas in autovia construction contracts; Ministerio de Obras Públicas y Transportes, "Informe Sobre el Primero Plan."

two-thirds of the increase was based on prices for construction materials (steel and petroleum) and only one-third on prices for operating inputs (labor). Henceforth, toll increases would be 95 percent of the consumer price index.

More significantly, the government began debating the possible use of toll autopistas for the 1993 highway plan, largely because of tight government budgets. The possibilities for private toll roads seemed limited, however, since the remaining unbuilt roads would probably not be profitable from tolls alone. Tolling the existing autovías also seemed out of the question, if only because the typical autovía was built on the old national road right-of-way so no untolled alternative is available. The major future opportunity therefore may be to adopt the French system of persuading the existing autopista companies to cross-subsidize new routes with profits on older routes.

The Merits of Toll Financing

The experiences of France and Spain suggest that a developed country can finance an extensive intercity highway network with tolls, but they provide mixed evidence on whether toll financing is superior to tax financing.

Financial Viability

Both France's 5,500-kilometer system of toll autoroutes and Spain's 2,000-kilometer system of toll autopistas appeared to be largely self-supporting in the early 1990s, although it would be difficult to develop an exact accounting of past subsidies received from the French and Spanish governments. In France, six of the eight autoroute companies were generating sufficient toll revenues in 1990 to cover their operating expenses and debt service and to make repayments on their advances from the state, as shown in table 8-9. The remaining two companies (ACOBA and AREA) had fairly small losses that were more than offset by the profits from their peers, so that the ADF clearinghouse was generating a significant surplus to retire old state advances.

The French autoroute system as a whole probably still would be profitable if a complete accounting of past subsidies were included. Autoroute concessions received interest-free cash advances covering about 30 percent to 40 percent of their construction costs until 1970, and the forgone interest is a subsidy not reflected on the ADF's books. (Cash advances were less generous and to be repaid with interest after 1970.) In

TABLE 8-9. Status of French Autoroute Concessionaires, December 1990

Company (principal routes)	Kilometers: In service	Kilometers: Con-ceded	Average traffic intensity (veh/day)	1990 Financial results (millions of francs): Total revenue	1990 Financial results (millions of francs): Profit (loss)	Debt, including ADF advances (millions of francs): Total debt	Debt, including ADF advances (millions of francs): Advance only
Autoroute companies							
ACOBA (Côte Basque)	66.5	66.5	14,102	204	(14)	1,762	890
AREA (Alpine autoroutes)	283	464	20,076	897	(63)	7,157	1,089
ASF (Marseille–Lyon and southwestern France)	1,500	1,635	23,506	4,822	1,051	20,998	2,851
ESCOTA (Marseille–Nice area)	368	428	28,549	1,682	61	11,228	1,950
SANEF (Paris–Lille and northern France)	926	1,004	18,582	2,382	278	13,576	3,465
SAPN (Paris–Caen)	187	214	25,277	559	274	723	0
SAPRR (Paris–Lyon and central France)	1,303	1,769	20,619	3,377	461	20,093	267
Cofiroute (Paris–Poitiers, Paris–Le Mans, Orléans–Bourges)	732	760	22,630	496	69	1,170	187
Tunnel companies							
Société du tunnel routier sous le Mont-Blanc (STMB)	117	117	13,636	531	194	1,465	83
Société Française du tunnel routier du Fréjus (SFTRF)	13	13	2,726	204	(9)	1,047	664

SOURCES: *1990 annual reports* of Autoroutes de France and Cofiroute, Paris.

addition, a share of their loans was guaranteed and in-kind advances were sometimes made. The cash advances for the earliest concessions probably could have been repaid by the late 1970s or early 1980s, however, if the national government had not decided to divert autoroute surpluses to support expansion of the network during the mid- to late 1970s. Even with network expansion, moreover, by the late 1980s the system as a whole was beginning to make such significant repayments to the state that it seems likely that forgone interest might eventually be recovered too.

Similarly, while only five of the thirteen Spanish autopista concessions were clearly profitable in 1990, these five alone accounted for 76 percent of the route kilometers in service and 89 percent of autopista traffic (table 8-10). Three of the five were highly lucrative (ACESA, AUMAR, and Europistas), while the other two (Iberpistas and AVASA) were already reasonably profitable and thought to have even greater profit potential as traffic grows.[53] Of the eight remaining autopista companies, three are so new that their profit potential has not really been tested (AUTEMA, AUCAT, and TABASA); two or three others might eventually prove profitable although their prospects are uncertain (Eurovias, AUDASA, and perhaps AUDENASA)[54]; while the remaining two (AUCALSA and Tunel del Cadi) seem to have few prospects.

The five profitable Spanish autopista companies benefited from past government assistance but probably still would be profitable if charged the value of these subsidies. The primary forms of government assistance have been guarantees of foreign loans and exchange rates. None of the five has drawn on the loan guarantees, although the guarantees were probably essential to win initial financing.[55] Four of the five have been the source of heavy exchange rate losses for the government, and government officials frequently note that the sum of the exchange losses over the years is five to six times the equity invested by autopista shareholders. This exchange rate insurance may have been worth far less to the companies than its cost to the government, however. The companies might not have paid significantly higher interest rates if they had looked for foreign loans with low exchange rate risk or been allowed unrestricted access to the domestic capital market.

The three autopista companies taken over by the national government in the mid-1980s (AUDASA, AUCALSA, and AUDENSA) have received far more substantial government aid than the rest, primarily in interest-

53. One sign of the profitability of these companies is that the shares of four of them (all but AVASA) have been listed and traded on the Spanish stock exchange since the late 1980s .

54. AUDASA might have been marginally profitable in 1990, for example, had its toll rates been closer to the industry average. Even so, these companies might never earn enough to compensate their investors for the many years without dividends.

55. The government charges the autopista companies twenty basis points for these guarantees, moreover. As a result many of the profitable companies no longer find the guarantee worthwhile; Ceron and Jerez, interviews.

TABLE 8-10. Status of the Spanish Concessionaires, 1990

Company (principal routes)	*Year begun*	*Route kilo-meters in service*	*Vehicle kilo-meters (thou-sands/ day)*	*Average traffic intensity (veh/day)*	*Average toll (pts/km)*	*Financial results (millions of pesetas)*	
						Total revenue	*Profit (loss)*
Private companies							
ACESA (Barcelona–France, Barcelona–Zaragoza)	1967	509	4,603	23,435	9.28	36,443	17,265
Europistas (Bilbao–France)	1968	106	632	16,386	15.09	9,652	3,837
Iberpistas (Northwest Madrid)	1968	70	356	14,005	14.67	5,268	526
AUMAR (Tarragonna–Alicante, Cadiz–Sevilla)	1969/71	468	1,914	11,214	12.62	21,866	9,651
AVASA (Bilbao–Zaragoza)	1973	294	738	6,873	14.62	10,031	1,452
Túnel del Cadí, (North of Barcelona)	1973	30	54	4,989	31.03	1,352	(1,718)
Eurovías (Burgos-Cantabrico)	1974	84	255	8,294	12.33	2,838	(276)
AUTEMA (Near Barcelona)	1987	33	82	6,730	15.94	967	(1,285)
AUCAT (Near Barcelona)	1989	0	...	...	...	...	...
Public companies							
AUDENASA (Tudela–Pamploma)	1972	99	182	5,041	10.22	1,481	(1,381)
AUDASA (Ferrol–Portugal)	1973	115	375	11,632	8.55	3,108	(1,838)
AUCALSA (Leon–Campomanes)	1975	87	108	3,820	14.16	1,546	(3,311)
TABASA (Vallvidrera tunnel)	1987	0	...	...	...	...	...

SOURCE: Ministerio de Obras Públicas y Transportes, *Memoria 1990*, pp. 7, 17, 25, 33, 91.

free construction advances and payments for operating deficits. These companies are the exception rather than the rule, however, accounting for only 16 percent of autopista route kilometers and 7 percent of vehicle kilometers in 1990.

This general picture of profitability must be qualified in several important respects. In the first place, experiences in both France and Spain suggest that an extensive and truly national system of intercity high-performance expressways probably could not be financed with tolls without the use of cross subsidies. The extent of cross subsidies in France is apparent in table 8-11, which shows the recent traffic volumes on different autoroute segments for three currently profitable companies. Industry observers estimate that toll revenues cover total operating and construction costs when daily traffic approaches 20,000 vehicle-kilometers per kilometer of autoroute, assuming typical construction costs and toll rates. Each of the three profitable companies averages around 20,000 vehicles a day over its entire network, but volumes vary from fewer than 10,000 vehicles on some of the most recently built autoroutes to as many as 40,000 to 50,000 vehicles on older segments. Moreover, some of the later autoroutes may never reach profitable traffic volumes simply because they are branches or secondary routes with less inherent traffic potential than the earliest trunk autoroutes. Cross subsidies among, as opposed to within, companies have declined since the early 1980s and were fairly unimportant in the late 1980s (as shown in table 8-6).

Spain chose to build half of its intercity expressway system with tolls and half without, but Spain probably would have had to use a system of cross subsidies similar to that of France had it decided to finance the entire network with tolls. Cross subsidies would seem especially necessary in Spain because Spanish traffic densities are much lower than those in France, owing in part to Spain's significantly lower population density and its slightly lower per capita income and auto ownership. In such circumstances, Spain's existing toll roads are profitable largely because its toll rates are approximately twice as high as in France. (In 1990 the average toll rate was 12 pesetas, or U.S. $0.12, per vehicle-kilometer in Spain versus 0.32 francs, or U.S. $0.06, per vehicle-kilometer in France.) At these toll rates, Spanish highway planners estimate that an autopista with typical construction costs will be profitable if it carries an average of 10,000 vehicles a day, a figure half as high as in France.[56] While much of Spain's toll autopista system carries traffic volumes above these levels, a significant share apparently does not.

A second important qualification is that the French and Spanish experience offers little guidance on the prospects for financing urban expressways through tolls. Both countries have built their urban expressways

56. Martin and Fernandez, interviews.

TABLE 8-11. Variation in Traffic Intensity, Selected French Companies and Routes in 1990 or 1989

Company and autoroute	Traffic intensity in vehicles per kilometer each day
SAPRR (1990)	
Beaune–Lyon	48,301
Paris–Beaune	32,249
Lyon to A-40	14,363
Macon (A-6) toward Mt Blanc	14,002
Beaune–Mulhouse	13,823
Beaune–Toul	13,644
Clermont-Ferrand–Bourges	8,360
Langres (A-31)–Troyes	3,884
Tunnel Ste Marie aux Mines	2,522
Network average	20,619
SANEF (1990)	
Paris–Lille	46,887
Paris–Strasbourg	15,097
From A-1 NE around Lille	14,987
Calais–Reims	8,537
Network average	19,993
Cofiroute (1989)	
Paris–Orléans	56,550
Orléans–Tours	26,896
Paris–Le Mans	23,906
Tours–Poitiers	19,433
Le Mans–Rennes	13,961
Angers–Nantes	10,582
Orléans–Bourges	10,421
Network average	21,050

SOURCES: SAPRR, *Rapport Annuel 1990* (Paris), p. 14; SANEF, *Highlights 1990* (Paris), p. 8; Association des Sociétés Françaises d'Autoroutes, Bulletin d'Information, no. 46 (1990), p. 25.

with tax financing, although France is now turning to tolls to complete some key missing segments and Spain is considering the idea. (Arguably, some of the recent concessions in Catalonia could be classified as urban, although they are on the outskirts.) Cross subsidies will not be available to build the new French urban toll autoroutes, however, since the existing urban autoroutes are untolled, and the French government has decided not to force the intercity SEMs to build urban roads (with a few recent exceptions near Paris and Lyon). It is unclear whether toll financing will prove feasible for many of the missing urban segments since they are typically the most environmentally sensitive and costly to construct, although they also have great traffic potential.

Finally, profitability clearly depends on the government's willingness to maintain reasonable toll rates, as France's experience clearly illustrates. The French Ministry of Finance's decision to hold down toll increases after the energy crisis of 1973–74 contributed to the industry's serious financial problems in the late 1970s and early 1980s, especially since the Ministry of Equipment was encouraging network expansion at the same time.

The Desirability of Toll Financing

The French and Spanish experience is somewhat equivocal on the question of whether tolls are a better means of financing high-performance roads than fuel taxes or other revenue sources. The objection that collecting tolls is very costly does not seem to be supported by experience in either country, at least for intercity roads. In France, for example, construction of toll plazas and related equipment adds on average about 10 percent to construction costs while operating expenses for toll collection account for only 10 percent to 12 percent of receipts.[57] Delays to motorists are probably fairly minimal as well, since the French and Spanish toll roads are intercity routes where relatively infrequent mainline toll barriers or closed ticket systems (with collection on exit) can be used. The French have been experimenting with electronic toll collection as well, to reduce collection costs and motorist inconvenience further. Even without these economies, toll collection costs would appear roughly comparable to or not much greater than the collection costs for most alternative revenue sources.

Spain and France seem to differ significantly, however, on the degree to which tolls contribute to congestion on parallel routes. French autoroute experts do not regard this congestion as a serious problem, since they estimate that tolls divert only 6 percent to 7 percent of potential autoroute users to parallel routes.[58] Spanish highway planners, however, believe that tolls have caused great misallocations of traffic between some autopistas and parallel untolled national routes. Between Cadiz and Sevilla and along the Mediterranean coast, for example, the autopistas are relatively underutilized while the parallel national roads are severely congested. This situation has led Spanish highway planners to speculate that it sometimes might be less expensive to relieve congestion by paying the autopista concessionaire to reduce tolls than by improving the parallel

57. Seligmann, "The French Highway System," p. 40.
58. Seligmann, "The French Highway System," pp. 44–45.

national road. The greater misallocation of traffic in Spain is probably because of the higher toll rates charged by Spanish concessionaires.

Toll financing appears likely to have improved the quality of investment decisions more markedly in Spain than in France. The first French autoroute master plan of 1960 was based on the premise that the profitability of individual segments was an appropriate guide for investment, since toll receipts reflected the value that the users received. Since the mid-1970s, however, France's autoroute construction program seems dictated largely by the availability of cross subsidies from older segments rather than by any comparison of the costs and benefits of new sections. One can argue, of course, that there are important social and economic advantages to having an extensive network of high-performance roads rather than a smaller network in which each individual segment is, or has the potential to be, self-supporting. Nevertheless, toll financing does not seem to have encouraged financial discipline or careful cost-benefit analysis in France.

By contrast, Spain's reluctance to use extensive cross subsidies seems to encourage investment discipline for both the toll-financed autopistas and the tax-financed autovías. For autopistas, the lack of cross subsidies forces careful consideration of the profitability of each potential concession. For autovías, the competition for limited government budget resources encourages similar caution (as Spanish highway planners discovered after their heavy investments of the late 1980s).

The Advantages of Private Concessionaires

France's experience suggests that private companies probably can build and operate roads more cheaply than public companies. The principal evidence for cost advantages comes from only one case, Cofiroute, however, which is not necessarily typical since Cofiroute is also the only remaining private concessionaire in France. Cofiroute's construction costs per kilometer for its initial 1971 concession were 23 percent below the comparable figures for SEMs at the time. Ten percentage points of the difference were reportedly because of more cost-sensitive roadway design, including a slightly narrower median and the use of swale drainage systems where possible. The remaining 13 percentage points were from higher productivity of labor and equipment.[59] Cofiroute is also reputed to

59. The projected costs are reported in Fayard, *Les Autoroutes*, p. 69; despite unanticipated inflation after the oil crisis of 1973–74 Cofiroute's actual expenses were only 1 percent to 2 percent higher than projected according to Cyna, interview.

be a relatively efficient autoroute operator, although this impression is difficult to confirm from simple comparisons of Cofiroute and SEM operating statistics since their autoroutes all differ significantly in ways that affect productivity and costs.[60] In the Spanish case there is also no easy way to test the hypothesis that private companies are more efficient at building or operating toll roads; there are only a few public companies in Spain, and most operate under unusually difficult circumstances (such as mountainous terrain and tunnels).

Experiences in both France and Spain point, however, to the danger of creating incentives under which private concessionaires are more interested in building than operating a toll road. The combination of low equity requirements, advances, loan guarantees, and foreign exchange insurance almost surely weakened the incentives of shareholders to control costs, especially since construction companies were major backers. That the former private owners of now-public companies in both countries agreed to build roads in unusually difficult circumstances suggests that they were not very concerned with operating profitability.

However, these problems may have been less serious in Spain than in France, perhaps because the government assistance in Spain was slightly less generous. None of the five private companies granted concessions in Spain during the late 1960s had to be taken over by the government. More telling, only three of the eight major private companies granted Spanish concessions during the early 1970s were taken over (although another two had to be absorbed by previously established private companies). By contrast, three of the four private companies established in France during that same period are now government owned.

It is unclear to what extent the apparent conflicts of interest or weakened incentives to control construction costs could have been avoided, particularly with the earliest toll roads in both countries. On the one hand, the blame could lie with the terms of the concessions rather than with the concept of privatization. On the other hand, high equity requirements, no state guarantees, or prohibitions on construction company shareholders may have made the earliest French and Spanish toll roads unfinanceable given the lack of a track record to reassure private investors. Even with later private concessions, some assistance may have been necessary, especially for more lightly traveled routes where profitability might remain highly uncertain. Nevertheless, some of the undesir-

60. See Gómez-Ibáñez and Meyer, *Toll Roads in France and Spain*, pp. 45–47 .

able incentives created may have been unnecessary, particularly the Spanish government's assumption of the full foreign exchange risk.

If private concessions do offer cost advantages, the French government no longer seems to take advantage of this possibility as a means to improve public enterprise performance. Concessions have not been awarded competitively since the early 1970s, and even then the SEMs did not compete directly against private companies. The government's current reliance on cross subsidies to finance new SEM concessions makes public-private competition more difficult but probably not impossible. The SEMs could be required to maintain separate accounts for their present networks and any new concessions, for example, and bids from both private companies and SEMs could be judged on the degree to which they include requests for advances from the ADF clearinghouse.

Experiences in both France and Spain also illustrate the dilemmas of regulating the toll rates of private concessions. France leaves toll increases to the discretion of the Ministry of Finance, which generally grants toll increases at less than the rate of inflation and in 1991 gave the private toll road operator smaller increases than the SEMs, on the grounds that the private company was more profitable.[61] By contrast, Spain has always used a formula for toll increases and has changed formulas only once (at the autopista industry's request). The uncertainty under the French system has clearly been an obstacle to new private investment. The tailoring of toll increases according to financial need may also reduce the incentives for French operators to be efficient, especially because the criteria for toll increases are not explicit. However, the simple industrywide formula used by Spain risks unnecessarily high returns to investors in the most profitable roads. Whether this problem has occurred in Spain is uncertain; a detailed comparison of the historic returns to autopista investors with the returns required by investors in other private projects facing similar risks would be required to do an evaluation. Some compromise between the French and Spanish approaches would seem advisable, however, particularly one that combined the predictability of a formula (or a well-understood set of rules) with the possibility of limiting toll increases on roads that were exceptionally lucrative.

61. The French Ministry of Finance only began to vary the toll increases according to the different companies' perceived financial needs with its August 1991 toll increase, the first increase granted since 1988. In its initial order, the ministry awarded the SEMs increases of 4 percent for light vehicles and 8 percent to 9 percent for heavy vehicles but gave Cofiroute no increase at all. After strong complaints, the ministry eventually allowed Cofiroute a 4 percent increase for heavy vehicles only. Interviews with Cyna and Lecomte.

Chapter 9. Regulatory and Financial Trade-Offs: The Developing Countries

ALTHOUGH DEVELOPING COUNTRIES have a shorter history of experimentation with private toll roads than the European pioneers, their experience is instructive because of their different environmental settings. Some of the differences are physical and economic. Traffic volumes or densities are generally lower in the developing world than in Europe, for example, because per capita incomes are usually lower than in Europe. (This is true even if one compares France or Spain in the late 1950s or 1960s with most developing countries in the 1980s or early 1990s.) Traffic growth is usually concentrated in the few largest cities, moreover, in part because the disparities in economic wealth between the cities and the countryside tend to be greater. Perhaps as a result of these income differences, the developing countries have been more willing than their European counterparts to consider private toll roads in urban areas.

Other differences are institutional and political. Domestic private capital markets are typically less developed in the third world, as are links to international capital markets. Moreover, the political or regulatory environment in which a private infrastructure project must operate may be perceived, rightly or wrongly, as more uncertain, or at the very least as less familiar to the international investor. These differences generally make the use of privatization to tap new capital markets both more desirable and more difficult in the developing than the developed world.

Southeast Asia

Developing countries, in light of their unique conditions, have tried different approaches to road privatization and encountered different problems than those faced by developing countries. The following examinations of private toll road programs in three Southeast Asian countries and in Mexico illustrate this point.

Malaysia

In Southeast Asia, Malaysia, Indonesia, and Thailand have experimented with private toll roads. The Malaysian government began building a publicly owned national intercity toll expressway system in the late 1970s.[1] The original plan, approved in 1978, called for building a 785-kilometer north-south road stretching the entire length of the country (from the Thai border to Singapore) within five years. Another 143 kilometers of side links were to be built as well, the most important being in the region of Kuala Lumpur, Malaysia's capital and largest city. A public highway authority was created in 1980 to finance the road with government-guaranteed loans from the private capital market.

Costs proved higher and construction slower than expected. The public highway authority was soon effectively bankrupt, especially since some of the earliest construction contracts were for unusually expensive or lightly traveled sections of the route. The World Bank rejected the Malaysian government's request for a loan to support the project because of concerns that a uniform four-lane expressway was not needed for the entire length and that tolling might discourage the utilization of the road unduly. By 1986 the government, strained by the losses on the sections already completed, decided to privatize the road in order to complete it.

In 1987 the government awarded a thirty-year concession to United Engineers of Malaysia (UEM), a newly created firm whose principal shareholders included senior government officials. Other private firms with a proven track record, and possibly stronger financial backing, had reportedly been interested in the project but were discouraged because the government initially refused sufficient guarantees and assurances. For UEM, however, the government was fairly generous. UEM would operate and retain the tolls on the 424 kilometers of roadway that had been built and opened already, while not assuming any responsibility for the M $3.1 billion (U.S. $1.2 billion at prevailing exchange rates) spent on building the completed segments. Toll rates were to be doubled (from M $0.025 to

1. This description of the Malaysian experience is based primarily on Maurice Le Blanc, "Toll Road Experience in Malaysia," pp. 142–47 in Frida Johansen, ed., *Earmarking, Road Funds and Toll Roads: A World Bank Symposium*, Report INV-45 (Washington: World Bank, June 1989); Frida Johansen, "Toll Road Characteristics and Toll Road Experience in Selected South East Asia Countries," *Transportation Research A*, vol. 23, no. 6 (1989), pp. 463–66; and Mark Augenblick and B. Scott Custer, Jr., *The Build, Operate, and Transfer ("BOT") Approach to Infrastructure Projects in Developing Countries*, Policy Research and External Affairs Working Paper, WPS 498 (Washington: World Bank, August 1990), annex, p. 8.

M $0.05 per vehicle-kilometer, or from U.S. $0.01 to $0.02 per kilometer) and would be linked to the consumer price index after 1995.[2] In addition, the government would lend the company M $750 million during construction and another M $950 million on completion. Traffic levels would be guaranteed by the government, and the government agreed to assume some of the foreign exchange and interest rate risks and to compensate UEM if the project were delayed or costs increased as a result of government policy.

Even with these terms, however, UEM found it difficult to raise the financing needed to complete the project (either on foreign or domestic capital markets). As the World Bank foresaw, the highway may have been so overdesigned as to be financially risky even with heavy government support.

Indonesia

Indonesia also began a public toll road program in the late 1970s, although on a somewhat more limited scale than Malaysia.[3] During the 1970s highway traffic grew rapidly throughout Indonesia, particularly in and near the major cities (Jakarta and Surabaya) of Java, Indonesia's most populous and developed island. The Indonesian government was committed to spreading the benefits of development widely, however, and thus planned to focus its limited road construction budget on building rural development roads in economically lagging areas and islands. If high-performance highways were needed in Java and its major cities, toll financing seemed appropriate.

A government toll road authority, Jasa Marga, was created in 1978 to implement the toll road program. Although Jasa Marga was to be financially independent and self-supporting, approximately two-thirds of its investments were financed through foreign loans on which the Ministry of Finance, rather than Jasa Marga, made all interest and principal payments.[4] By 1990 Jasa Marga had built 318 kilometers of toll road and four bridges. The most heavily traveled routes included several short

2. The United Engineers had originally requested that it be allowed to raise tolls to M $0.075, but public outrage forced the government to roll back the increase to M $0.05. The government lengthened the company's concession from twenty-five to thirty years in partial compensation; and Le Blanc, "Toll Road Experience in Malaysia," p. 145.

3. This account of the Indonesian experience is based in part on Johansen, "Toll Road Experience in Selected South East Asian Cities," pp. 465–66; Sweroad, *Update of Toll Road Investment Program Study*, report to the Republic of Indonesia (Jakarta: Jasa Marga, April 1990); and interviews by the authors in Indonesia in June 1992.

4. The remaining one-third has been financed primarily through bonds sold to government employee pension funds on which Jasa Marga pays slightly below-market interest.

urban toll roads in the Jakarta metropolitan area and several routes extending from Jakarta out twenty-seven to seventy-five kilometers into the surrounding countryside.

The Indonesian government became interested in private toll roads in the late 1980s. Indonesia, an oil exporter, suffered economically from the fall in real oil prices at the beginning of the decade. By the late 1980s the Indonesian economy was showing signs of recovery and rapid traffic growth was expected to resume. Jasa Marga's resources had been strained by the construction undertaken in the 1980s, however, since these projects included some very expensive urban expressways and relatively lightly traveled intercity routes that were less profitable than earlier roads. Under Indonesian law only Jasa Marga can own and operate a toll road, but private companies could be involved in joint ventures with Jasa Marga. Indonesian law also mandated that the toll road right-of-way be owned by the government, which was interpreted to require that the Ministry of Transport and Communications pay for the cost of assembling and clearing the right-of-way. The government therefore proposed joint venture schemes in which construction costs, but not right-of-way acquisition, were to be financed primarily by private equity and debt, with Jasa Marga contributing a small part of the equity. A list of twenty-two proposed private toll roads was circulated by the government to potentially interested private investors, and an interministerial team was formed to organize and evaluate competitive proposals for those segments that elicited significant private interest.

By 1992 two private joint venture toll roads were opened, two more were under construction, and the government was in the process of negotiating final agreements with selected joint venture consortia or selecting from among competing joint venture proposals for approximately a dozen other roads. The roads opened or under construction were awarded to consortia that included firms owned by politically prominent citizens or government officials, as had been the case in Malaysia. Substantial government assistance was received in a variety of forms (beyond government payment for right-of-way). Virtually all of the debt was supplied by the national government's development bank (BAPINDO) or by other government-owned banks. A substantial share of the equity, between 25 percent and 75 percent, was supplied by Jasa Marga, often in the form of key interchanges, connecting bridges, or short road segments it had already built. Jasa Marga's contributions also appear to have been undervalued, primarily by using historic rather than current costs. Most of the remaining equity is "sweat equity," coming from contractors and engineering firms in the concession consortia rather than cash. Finally, the two roads opened have rather generous toll revenue-sharing agreements with adjacent or connecting Jasa Marga roads.

These arrangements were partly because of the time pressures on the first joint venture negotiations and partly because of the weakness of Indonesia's private domestic capital market. The government had declared 1991 as "Visit Indonesia Year" and wanted certain key infrastructure projects, including the first two roads, completed before the year began. The domestic private capital market is also very small and underdeveloped in Indonesia. Debt of more than a few years' duration is rare and hard to place in the private market, especially in the amounts required by toll roads, which left government banks and pension funds as the only viable domestic source.

Private foreign investment in Indonesia's toll roads has also not been readily forthcoming because Indonesia's toll road law specifies that all toll increases are at the discretion of the president. Consortia including private foreign investors have been selected as joint venture partners for several toll roads, but the government's unwillingness to amend the law or otherwise guarantee a formula for future increases in toll rates has forced at least one of these foreign-led consortia to withdraw and has blocked final agreements with others. The government has been unwilling to commit itself irrevocably to a formula because it is unsure how lucrative these projects might be and wants to retain the possibility of slowing toll rate increases if profits appear excessive. Without such assurances, it is not surprising that only politically well-connected investors would consider becoming involved in toll road joint ventures.

Indonesia's private toll road program has thus been largely financed by government-controlled domestic sources of capital, rather than new private domestic or foreign funds. Broadening the sources of capital on which it draws may be important in the future, since Indonesia's toll road program is fairly ambitious.

Thailand

Thailand's private toll road program has focused on urban expressways in Bangkok rather than on intercity roads,[5] thus running somewhat

5. This account of Thailand's experience is based partly on Johansen, "Toll Road Experience in Selected South East Asia Countries," p. 466; Joseph W. Ferrigno III, "The Bangkok Second Stage Expressway Project," *Public Works Financing*, July 1990, pp. 16–17; "Tolled Ring Road Led the Way for Bangkok BOT Projects," *Public Works Financing International*, July 1992, pp. 4–7; and Augenblick and Custer, *Build, Operate, and Transfer*, annex, pp. 8–10.

counter to the usual experience. The government has upgraded and imposed tolls on several intercity roads since the 1970s and has attempted to attract private investment into intercity toll roads since the late 1980s, but without notable success.

In Bangkok, the government established the Expressway and Rapid Transit Authority (ETA) in 1972 as a state enterprise to implement toll expressways and rail mass transit systems. ETA took more than a decade to build Bangkok's twenty-seven-kilometer first-stage toll expressway (completed in 1984). Bangkok's economy and traffic congestion were growing so rapidly by then that government officials feared traffic congestion would soon choke off Bangkok's economic growth unless the pace was accelerated. The government began to consider private road and rail schemes, including private provision of planned second-, third-, and fourth-stage expressways.

In 1989 the government signed a thirty-year concession for the second-stage expressway with Bangkok Expressway Company Limited (BECL), a Thai company created and majority owned by Kumagai Gumi, the giant Japanese construction and engineering firm. This concession was apparently not awarded competitively[6] but involved significantly more private financing than the early Malaysian or Indonesian concessions. The 27.5 billion baht (U.S. $1.1 billion) construction costs were financed 20 percent by equity and 80 percent by debt. Two-thirds of the initial equity would be supplied by Kumagai Gumi and the remaining one-third by Thai investors, mainly banks. Most of the debt was raised on Thai domestic capital markets, but a portion was guaranteed by international banks.[7] Kumagai Gumi provided warranties to BECL and the government that the project would be completed within budget and on schedule (by 1995).[8]

Critical to the agreement, however, were government assurances on tolls and land assembly. Together with ETA's first-stage expressway, the thirty-kilometer second stage would complete an inner-ring road around central Bangkok with spokes to the north, east, southwest, and southeast. Toll collection on the first and second stages was to be integrated to avoid inconveniencing motorists unduly. BECL and ETA agreed on a formula

6. A second bidder was disqualified for lack of experience, although some observers regarded the BECL as having an inside track from the start.

7. Approximately U.S. $250 million; Ferringo, "The Bangkok Second Stage Expressway."

8. The first segment of the second-stage expressway is to open in 1993, and the project to be completed in 1995.

for sharing tolls on the two expressways that gives BECL the larger share in early years.[9] Toll rates are also to be tripled when the second stage is opened and indexed to inflation at five-year intervals thereafter.[10] ETA is responsible for acquiring the land on schedule and paying the 10 billion baht (U.S. $400 million) land acquisition costs, although BECL is to pay ETA 16 billion baht in land lease fees over the life of the project. The agreement also provides for remedies, such as toll increases, should the government delay the project or other adverse circumstances develop.

Mexico's Private Toll Road Program

Mexico has shifted between toll and tax financing of its high-performance highways several times as the resources and fortunes of its public sector have waxed and waned. Mexico began building its high-performance road system as publicly owned toll roads in the 1950s. Approximately 1,000 kilometers of public toll road were eventually opened, with most construction taking place before 1970.[11] These roads are mostly operated by the federal toll road authority, CAPUFE (Caminos y Puentes Federales de Ingresso y Servicios Conexos), which is an agency of the Ministry of Communications and Transportation. The public toll roads are concentrated around Mexico City, where traffic volumes are generally highest.

In the early 1970s, however, Mexico shifted to tax financing for new expressways, fueled by the propublic sector regimes of Presidents Luis Echeverria Alvarez (1970–76) and Jose Lopez Portillo (1976–82). By the mid-1980s, almost 3,000 kilometers of untolled four-lane divided highways had been opened to supplement CAPUFE's 1,000-kilometer toll network. Many of these highways were near Mexico City too, but

9. BECL and ETA will share toll receipts 60/40 in the first years of the BECL franchise, 50/50 in the middle years, and 40/60 in the final years. Ferringo, "The Bangkok Second Stage Expressway," p. 17.

10. Tolls on the first stage expressway were 10 baht (U.S. $0.40) in 1990 and will be raised to 30 baht at the start of the operating period for the second-stage expressway, in 1993. Ferringo, "The Bangkok Second Stage Expressway," p. 17.

11. During the 1950s, 226 kilometers were opened, and during the 1960s another 640 kilometers; only about 80 kilometers were opened after 1970. For an account of the early Mexican toll road system see Robert N. Panfield, "Toll Highways in Mexico," pp. 157–70 in Johansen, *Earmarking, Road Funds and Toll Roads;* and Gabriel Castaneda Gallardo, "Financing, Construction and Operation of Highways through Concessionary Arrangements in Mexico," pp. 171–90, in Johansen, *Earmarking, Road Funds and Toll Roads.*

substantial segments were built in the north around other major cities and on routes to the border with the United States.

The rapid expansion of the public sector combined with the collapse of oil prices in the early 1980s adversely affected the Mexican economy, which led to another dramatic reversal of government polices. With the Mexican economy experiencing negative growth rates in 1982 and 1983, government budget deficits grew to as much as 16 percent of the gross domestic product, and the financing of these deficits helped stimulate severe inflation. In response, President Miguel de la Madrid (1982–88) initiated a program to cut the size of the public sector, which his successor, President Carlos Salinas de Gortari (1988-94), expanded and accelerated. Privatization played a key role in these two presidents' strategies for reducing the government budget deficit, largely through the sale or liquidation of money-losing state enterprises.

Highways did not become a focus of the privatization program until the late 1980s. President de la Madrid raised tolls on the CAPUFE network beginning in 1982, but with the intention of diverting the proceeds to cover the government's budget deficit rather than to expand CAPUFE's system.[12] The recession also cut traffic growth rates and made road investments a less pressing priority. The need for road improvements became more apparent as the economy began to show signs of recovery, however, and in 1986 the de la Madrid government asked the national development bank, BANOBRAS, to study the possibility that new toll roads could be built as private concessions.[13] BANOBRAS was optimistic but recommended an experimental program to test the feasibility. Accordingly, two road concessions totaling 215 kilometers were granted at the end of the de la Madrid administration with BANOBRAS serving as the concessionaire and financing 50 percent of the project costs, the contractors financing 25 percent, and state governments the remaining 25 percent.[14]

President Salinas's Program

In February 1989, less than three months after taking office, President Salinas announced a dramatic new program to build 4,000 kilometers of

12. At this time CAPUFE's financial autonomy was ended, so that the government could divert CAPUFE surpluses to nonhighway uses; Panfield, "Toll Highways in Mexico," p. 159.

13. Castaneda, "Concessionary Arrangements in Mexico," pp. 171–72.

14. The two BANOBRAS concessions were Guadalajara-Colima (in the state of Jalisco) and Atalcomulco-Maravatio (in the state of Michoacan). A third concession (Tepic-San Blas) was awarded in its entirety to the government of the state of Nayarit.

new toll roads and seven new international toll bridges as private concessions before the end of his administration in 1994.[15] The projected cost was 11.5 trillion pesos (almost U.S. $5 billion at then-prevailing exchange rates), which tripled the rate of new highway investment over that of the preceding few years.

Accelerated road construction was viewed as an important part of President Salinas's effort to revive the Mexican economy. Road building would stimulate the economy by putting Mexico's construction industry back to work, and high-quality infrastructure was deemed critical to Mexico's long-term prospects. Private toll roads seemed the only choice, moreover, given that the government was trying to cut the public deficit and was in the midst of difficult renegotiations of the enormous foreign debts assumed during the public expansions of the 1970s.

Under the new program, the Ministry of Communications and Transportation would select the roads to be offered for concessions and specify the initial toll rates to be charged. Toll increases would be allowed to keep pace with the consumer price index. Bidders would be supplied with preliminary designs, cost estimates, and traffic projections prepared by the ministry. The concession would be awarded to the bidder that offered the shortest concession period, which could in no case exceed twenty years.

The new program would rely primarily on private concessionaires and financing, unlike the pilot projects of the previous administration. The concessions were to be awarded to consortia of construction companies and banks. The construction companies were expected to put up 25 percent to 30 percent of the cost in the form of "sweat equity" by discounting their construction bills by an agreed-on percentage. The companies could afford such discounts, the government reasoned, by deferring their normal profits and the depreciation on their construction equipment. The banks would finance the remaining 70 percent to 75 percent. To guard against conflicts of interest, each concession would create a special independent "trust" to review the contractor's bills,

15. For overviews of President Salinas's program see Mitch Stanfield, "Modernizing Highway Infrastructure through Toll Concessions in Mexico," report prepared for the Department of Transportation, Federal Highway Administration, Washington, December 1991; Victor M. Mahbub M., "Mexico's Private Road Concessions," *Public Works Financing International*, June 1991, pp. 1–5; and William G. Reinhardt, "SCT Considers Program Changes to Boost Capital for Concessions" and "Short-Term Concessions Skew Risk, Reward, Pricing in Contractors' Favor," *Public Works Financing International*, June 1991, pp. 6–11.

disburse the bank financing, and distribute the toll proceeds to the investors.

The government would guarantee its traffic and cost estimates in part. If traffic were less than the ministry forecast, the concessionaire could request an extension of the term of the concession. The concessionaire was responsible for the first 15 percent of any construction cost overrun; overruns in excess of 15 percent and any overruns caused by government-imposed delays or design modifications were grounds for requests for concession extensions. Direct public assistance for the concessionaires was to be kept to a minimum, except that the ministry would assemble the required right-of-way and lease it to the concessionaire for a nominal charge.

The emphasis on short concessions was dictated largely by the need to attract private capital. Financing long-term debt was nearly impossible in the Mexican domestic capital market during the 1980s, given the virulent inflation at the time; even medium-term (five-year) instruments were rare and could be sold only by the largest and most secure companies. The banks involved in the consortia preferred shorter concessions since they were reluctant to tie up their own funds for long periods. The contractors were also anxious to recover their sweat equity quickly, especially since some would be in the form of deferred depreciation on their equipment.

Mexican government officials were also concerned that the process of concession bidding be competitive and fair, and this concern favored making awards on only one criterion, such as the concession duration. The three pilot roads had not been awarded competitively by the de la Madrid administration, but competition may have seemed less important at the time since the concessionaires were essentially public agencies. Under the new program, competing proposals would differ only in the duration of the concession, since all bidders would be required to accept the ministry's route and toll rates. Competition along one dimension would make the process transparent to all and less subject to charges of manipulation or fraud.[16]

Accomplishments, Problems, and Modifications

By early 1992, the government had awarded 3,600 kilometers of these new toll road concessions and several for new bridges as well; approxi-

16. Secreteria de Comunicaciones y Transportes, "Programa de Carreteras y Puentes de Cuota." Carlos Mier y Teran O., coordinator of the Working Group, Mexico City, October 19, 1988, pp. 26–28.

mately 1,300 kilometers of the roads and two bridges were already opened for service.[17] The government's goal had expanded as well, to awarding at least 5,000 kilometers of new toll road and eight new bridges by the end of the Salinas administration (in 1994). The private concessionaires included most of the major construction companies and banks in Mexico. Mexico's three largest construction companies were awarded nine of the twenty-nine new concessions signed by the Salinas administration from 1989 through 1991, a number of smaller construction companies were awarded fourteen concessions, while state governments were granted six (these last were not awarded competitively).[18]

Mexico's pace of new toll road construction has been extraordinary. In opening approximately 500 kilometers a year in 1990 and 1991, Mexico rivals the construction rates in the peak years of the toll expressway programs in France and Spain. Not surprisingly, the Mexican government also encountered some problems along the way.

Cost and traffic projections. One such problem was inaccurate cost and traffic projections by the Ministry of Communications and Transportation. While government officials report that the costs on most projects were within 15 percent of the estimate, a number of dramatic underestimates occurred.[19] The Mexican government and the concessionaires blame the poor cost projections largely on the speed with which the concessions were awarded. The designs for the roads were often incomplete when the concessions were put out to bid. Sometimes the federal government or local communities pressed for modifications in the route after the concessions were awarded. Indeed, incomplete plans were somewhat advantageous in that they allowed the concessionaires more freedom to suggest changes in alignment or design when they encountered unexpected problems.

The errors in the traffic projections were blamed not just on the pace of the program but also on the high toll rates for the new concessions. In 1988, before the concession program started, CAPUFE tolls averaged

17. These figures apparently include the 240 kilometers of concessions granted in the last year of the de la Madrid administration; figures supplied by Andres Caso Lombardo, speech to the U.S./Mexico Roundtable on Concessionary Transportation Infrastructure, Manzanillo, Mexico, March 11, 1992.

18. These figures exclude the three concessions awarded under the de la Madrid administration; see Stanfield, "Toll Concessions in Mexico," pp. 27–29.

19. The government has not reported comparisons of actual and projected forecasts; the following account is based largely on unpublished speeches made by public officials and concessionaires and private interviews by the authors at the U.S./Mexico Roundtable on Concessionary Transportation Infrastructure, Manzanillo, Mexico, 1992.

only 57 pesos (then slightly over U.S. $0.02) per vehicle-kilometer;[20] however, the toll rates on the new concessions averaged 508 pesos (U.S. $0.17) per vehicle-kilometer in 1991.[21] Mexican highway planners had little experience with such high tolls and were sometimes surprised by the amount of traffic diverted to parallel untolled roads.

Inaccurate cost and traffic projections may also have been encouraged by the opportunity to renegotiate concession terms in the event of overruns. Concessionaires might have been more inclined to overlook inaccuracies in ministry forecasts when the concessions were being awarded, for example, because they understood that the concession terms might be renegotiated eventually to reflect more realistic cost and traffic figures. Moreover, some banks and other observers fear that the contractors had strong incentives to pad their construction bills to reduce the real equity that they contributed.[22]

Just how serious these incentive problems were is unclear. The government and the contractors generally deny that there were serious problems with cost padding, and some contractors argue that the government has been very tough in granting extensions in response to cost overruns. Obviously, however, bill padding and the opportunity for renegotiation could undermine the competitive procurement process, particularly the possibility of securing the lowest cost or most efficient contractor.

Short concessions, high tolls. The government has been successful in keeping concession periods short but at the cost of high tolls and potentially serious underutilization of some roads. The average duration of the first twenty-two private awards was only eleven years and ten months, with two of the awards for as little as five years and a few months.[23] These concession periods are especially short because they start as soon as the concession agreement is signed and thus include the time for construction as well as time to operate. Toll rates were correspondingly high, as noted earlier, and there were frequent reports that some of the new roads were underutilized while parallel roads remained congested. Tolls were so high on a road built to bypass the city of San Luis Potosi, for example, that most truckers reportedly continued to drive through the congested city center in order to avoid the toll.

20. Panfield, "Toll Highways in Mexico," pp. 160–61.

21. Stanfield, "Toll Concessions in Mexico," p. 11.

22. Stanfield, "Toll Concessions in Mexico," p. 14.

23. Calculated by the authors from figures in Stanfield, "Toll Concessions in Mexico," pp. 27–29.

The ministry seems to have set toll rates unnecessarily high in some cases, even allowing for its desire to reduce financing problems by keeping the concessions short. The ministry's ostensible objective was to set the lowest toll rates that would allow a financially viable concession of less than twenty years. By this standard it overestimated the tolls required, at least if the terms of the initial concession awards are to be believed. As in Spain, it is conceivable that motorists on some of these toll roads are now so price sensitive that toll rates could be cut without reducing total toll receipts or extending the length of the concession. By 1992 the ministry began studying the possibility of negotiating reductions in toll rates with some of the existing concessionaires in return for extensions of the concession if necessary.

Network expansion and profitability. The Mexican government awarded the most profitable concessions first, in keeping with its desire to limit public assistance. As in France and Spain, however, the profitability of the new awards declined as the concessionary network expanded, and Mexico was forced to gradually increase the direct government aid it offered for concessions. The primary source of the government aid has been surpluses generated by the existing CAPUFE toll roads.[24] Thus Mexico in effect began to adopt a scheme for cross subsidies similar to France's, except that the cross subsidies were among companies (from CAPUFE to some of the most recent private concessions) rather than within companies (as is largely the case in France).

Beginning in the spring of 1990, the ministry began to solicit bids for concessions that included government promises to pay a fixed share of the estimated construction costs. The government announced that its share of construction costs would never exceed 25 percent of total concessionary toll road investments, but that limit was soon reached. The concessions awarded through July 1990 were financed 29 percent by contractor equity, 61 percent by banks, 5 percent by the federal government, and 5 percent by state governments.[25] By early 1992, however, the

24. In 1992 the government also began to cross subsidize new private toll roads with proceeds from older private toll roads. It allowed one existing private toll road to extend the length of its concession in return for a payment (out of refinancing proceeds) that the government intends to use to fund its own equity contribution to a new private toll road; see "Lehman Rides High on Toluca Bonds," *Public Works Financing*, (July–August 1992), pp. 1–3.

25. Secretaria de Comunicaciones y Transportes, Subsecretaria de Infraestructura, "Attenta Nota Informativa Sobre el Apoyo de Caminos y Puentes Federales de Ingressos y SC en Algunos Grandes Proyectos del Programa de Autopistas Concesionadas," presentation to the Comisión Gasto-Financiamiento, Mexico City, August 1990.

contractor and bank shares had declined to 28 percent and 49 percent, respectively, while government's share, most of which was federal, had increased to 23 percent.[26]

The increasing burden of government assistance encouraged the Ministry to identify a priority national network of five main routes (figure 9-1). Three of these priority routes run from Mexico City north to the U.S. border. The other two routes extend from the Pacific to the Atlantic coasts and from Mexico City southeast to Veracruz and the Guatemalan border. Many segments of these routes had already been modernized by CAPUFE as untolled divided highways or had been among the early private concessions. The unimproved links would be awarded as new concessions, with government support as needed.

Interestingly, some segments in the planned national network are unprofitable not only because of low traffic but because of strong competition from untolled alternatives. This is especially true in the deserts of northern Mexico, where the existing national road is often straight, flat, and not congested by local traffic.

The government announced its intention to alter the form of public support for new concessions in early 1992. Future government contributions toward construction costs would be treated as equity investments instead of grants; they would thus earn the same returns as the sweat equity investments of the contractors, instead of no returns at all. The maximum government share would also increase, with 30 percent contractor equity, 40 percent government equity, and 30 percent debt being the suggested model.[27]

Limits on private capital. The most interesting aspect of Mexico's aggressive private road program is the way it has forced the Mexican capital markets and the government to devise new financial instruments to tap additional sources of funds. The Mexican banks have been innovative, probably because they are the most exposed. Contractors appear less strained by their equity requirements, perhaps because the possibility of padded construction bills has meant that their real investment was less substantial than claimed. The loans advanced by the banks clearly represent real risks, however. Many of the banks were still nationalized when the Salinas concession program began, and they may have been pressured by the government to participate initially.

26. Speech by Carlos Guash Cano, advisor to the Secretary, Ministry of Communications and Transportation, Manzanillo, Mexico, 1992.

27. Guash, speech.

FIGURE 9-1. Mexico's Priority National Highway Network

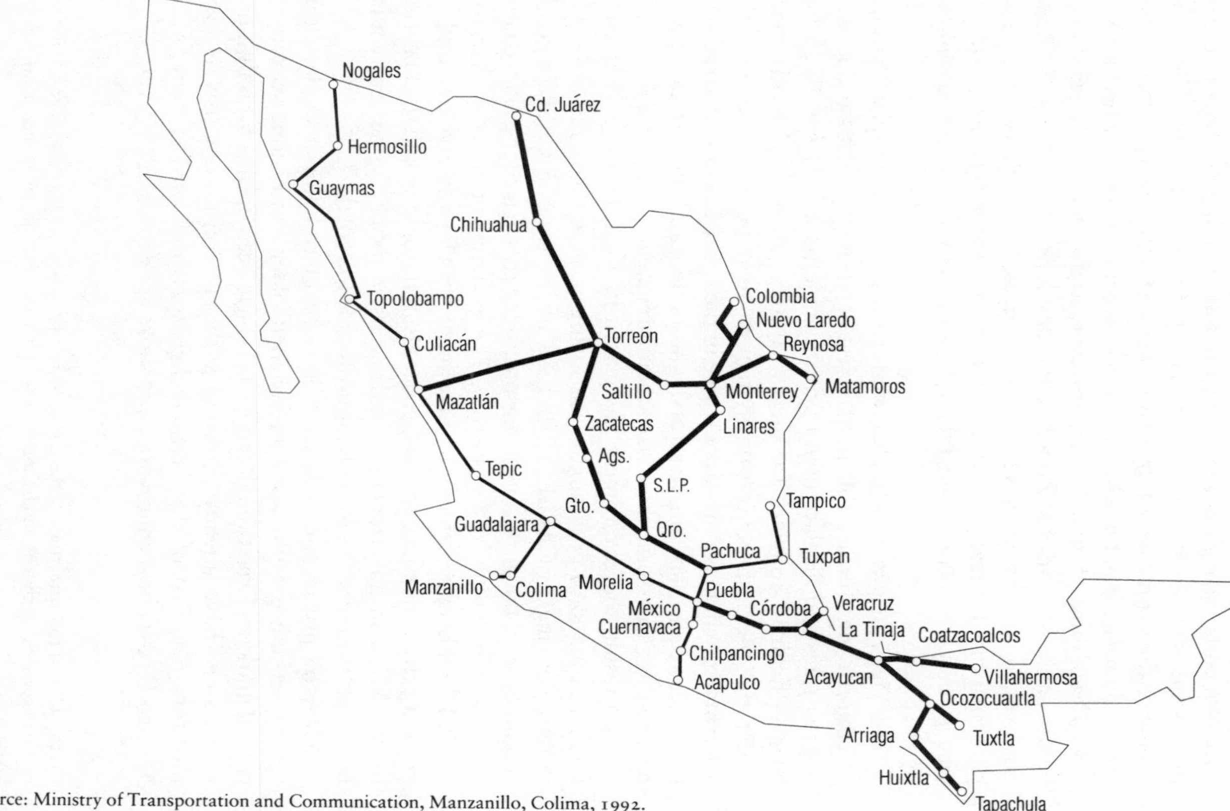

source: Ministry of Transportation and Communication, Manzanillo, Colima, 1992.

The banks have gradually broadened the pool of domestic investors involved in toll roads. Initially, most banks financed their share of construction costs through normal construction or commercial loans, drawing on the banks' existing pool of savers. Later many banks began to refinance their contribution by issuing medium-term infrastructure bonds on the domestic bond market; toll road revenues are not sufficient security to back most of these bonds, however, so they have been guaranteed by the bank. As roads have opened and developed a reliable traffic base, however, a few banks have begun to successfully sell certificates of participation that carry a fixed interest rate (over inflation) and are secured only by a claim against the toll road revenues and not guaranteed by the bank.

The Mexican government also wants to attract foreign capital, because of concern that further toll road investments might increase domestic interest rates and displace useful private domestic investment. But it has been difficult to convince foreigners to invest in *new* toll roads, since under the Mexican concession system there is no assurance that the government will make investors whole in the event of a cost overrun or a traffic shortfall. Mexican investors, who understand the system, appear more willing to assume these risks than foreigners.

As a result, the government has placed its main hopes for attracting foreign capital on a program of securitizing existing toll roads, especially those belonging to CAPUFE. The risks of investing in CAPUFE roads should be low, the government reasons, since they have long and reliable traffic and maintenance cost history. The government hopes to privatize CAPUFE with 30 percent equity investment by Mexican investors and 70 percent private foreign debt. The proceeds of the debt would then be used by the Mexican government to finance its proposed 40 percent equity investments in new but initially unprofitable toll roads.[28]

Mexico's private concessionaires have begun to follow a similar strategy by seeking to refinance some of their older concessions on foreign capital markets. Even though these roads have been opened for only a few years, some have already attracted sufficient traffic to satisfy foreign investors. The first of these refinancings was completed in 1992, for a Mexico City–Toluca concession, and more are expected.[29]

28. One of the attractions of the government's proposed 30/70 equity-debt scheme for new concessions is that it might make new toll road debt more attractive to foreign investors.

29. "Lehman Rides High on Toluca Bonds."

Capital Markets and Regulatory Environments

The experience of the developing countries illustrates the close relationship between two chief issues in infrastructure privatization: the desire to tap private capital markets and the need to balance investor and user concerns in designing regulatory schemes. Indeed, developing countries pose a paradox. On the one hand, the argument that privatization might increase overall investment in the economy by providing access to new capital is more credible in developing countries, especially since their capital markets are typically less sophisticated and integrated with world capital markets. On the other hand, the regulatory environment is often uncertain and risky in developing countries, making it difficult to attract new investors, especially from outside.

Attracting New Capital

Private domestic capital markets in developing countries may offer few outlets for long-term investments, especially when the country has a history of inflation or, worse, political instability. Even if the domestic capital market is reasonably stable and well developed, access to international capital markets may be limited. International investors may be reluctant to invest without government guarantees, for example, simply because of their inability to assess local conditions or their lack of confidence that the local legal system will safeguard their investment.

In this context, the traditional objection that private investment in highways (or other infrastructure) will simply displace other forms of private investment may be less compelling, especially where private infrastructure projects tap into foreign capital markets not previously accessible. But even if the private initiative drew on the domestic capital market only, total domestic investment still might be increased if privatization encouraged the development of new investment instruments that encouraged savings or less capital flight.

Of the four countries examined in this chapter, Malaysia and Indonesia seem least successful at increasing road investment without displacing other forms of investment. Malaysia's concessionaire for the north-south highway, UEM, is relying primarily on loans either directly from or guaranteed by the government; since the government has limited credit on domestic and international capital markets, UEM road investments probably come at the expense of some other government-financed or guaranteed investment. Similarly, Indonesia's joint ventures have relied primarily on government-owned banks, thus displacing other forms of domestic public and private investment.

Thailand and Mexico seem to have done better than Malaysia and Indonesia, although their record is mixed. In Thailand, Kumagai Gumi's equity in Bangkok's second-stage expressway represents an investment that would not have been made in the Thai economy without privatization, and the international bank guarantees may have encouraged some Thai domestic savings that otherwise would not have occurred. Otherwise, however, the second-stage expressway is to be financed on the domestic capital market. Mexico is the most successful, particularly since its planned program of privatizing CAPUFE and other existing toll roads is beginning to attract new foreign private investment to the country. The Mexican road privatization program may also have strengthened the domestic capital market and savings by encouraging the banks to develop new medium-term investment instruments, such as bank infrastructure bonds or toll road certificates of participation.[30] Even so, the bulk of Mexican private road capital, at least initially, probably came at the expense of other forms of private domestic investment.

Regulatory Environments

Success in tapping new capital markets is directly related to governments' willingness or abilities to fashion agreements that provide investors with reasonable security. Political or regulatory risk is as much an issue as business risk. Most investors, domestic or foreign, will want guarantees about key issues such as the rates of future toll increases and reasonable assurances that the government will not alter the terms of the franchise unilaterally. Where construction cost or traffic risk is great, investors will want the possibility of earning appropriately high rates of return, government guarantees, or both. The need for assurances seems greater for foreign than domestic investors, moreover, perhaps because they feel their political risks are greater, or at least more difficult to assess.

The more limited the government assurances, the more limited the pool of investors. The Indonesian government has been reluctant to commit itself to a toll formula, for example, so it is not surprising that only politically well-connected investors and government-owned banks

30. The development of these medium-term instruments would not have been possible, however, had not the government's economic policies reduced inflation and increased investor and saver confidence.

feel comfortable participating. Even in Mexico, where the government has agreed to a specific toll formula, uncertainty about government policy toward cost overruns and traffic shortfalls limited the initial road privatizations mainly to domestic entrepreneurs.

This problem is not restricted to developing countries, of course. France's Ministry of Finance appears to have inhibited domestic investment in private toll roads, for example, by disavowing the toll formulas specified in the private concession agreements of the 1970s and by refusing to specify a new toll formula.

In general, however, developing countries probably face special burdens because their political and regulatory traditions are not as well developed or understood, especially by foreign investors, as those of developed countries. Thus the places where access to capital markets might be most improved or enhanced through privatization are also the places where the regulatory and political environment makes tapping new markets exceptionally challenging.

Chapter 10. Private Toll Roads in the United States

INTEREST IN PRIVATE TOLL ROADS grew rapidly in the United States during the 1980s, although the United States still had far less modern experience with such initiatives than Europe or the developing countries. By 1993 state authorities had given preliminary approval to one private road proposal in Virginia and four in California, but construction had begun on only one of the projects. Several other states had passed or were considering legislation authorizing private road projects as well. Nevertheless, virtually all of the U.S. highway system as of the early 1990s was publicly owned and operated, and most of it was financed with motor fuel taxes or similar excises rather than tolls.[1]

The Development of the U.S. Expressway System

Private toll roads, however, are not without historical precedent in the United States. From approximately 1790 to 1850, the private toll road was the dominant form of overland transportation in the United States, and at least 10,000 and perhaps as many as 20,000 miles of private toll roads were in operation at the height of the turnpike boom.[2] The public roads that were built, moreover, were usually not built as competitors to

1. Most of the toll-financed facilities were, as explained below, public toll roads built before the 1960s. The few private tolled facilities included approximately a dozen bridges over major rivers in rural areas; these are either special development promotions of local business groups or remnants of the more extensive private toll road system of the eighteenth and early nineteenth centuries. Other exceptions included private toll roads inside important resorts built by a single developer, such as "Seventeen Mile Drive" in the Carmel, California, development owned by Del Monte properties or the road in the Avery Island Resort in Louisiana. Federal Highway Administration, *Toll Facilities in the United States: Bridges, Roads, Tunnels and Ferries* (Department of Transportation, April 1987).

2. John B. Rae, *The Road and the Car in American Life* (MIT Press, 1971), p. 18. Rae's excellent survey has been an important source for much of the historical discussion in this chapter. See also Daniel B. Klein, "The Voluntary Provision of Public Goods? The Turnpike Companies of Early America," *Economic Inquiry*, vol. 28 (October 1990), pp. 788–812.

the private turnpike system but rather as complements or supplements that better knitted the various private efforts into a more comprehensive and integrated network or system. The early nineteenth-century boom in private turnpike construction and public road building was halted, however, by the emergence of the railroad and, to a somewhat lesser extent, of canals.

The Beginnings of Modern Highways

The emergence of the internal combustion engine at the end of the nineteenth century revived the interest in highway building. This rebirth first focused on local farm-to-market roads, however, which were unlikely to generate sufficient traffic to pay for themselves through tolls. Accordingly, improvements were subsidized in various ways by tax revenues with construction undertaken by state or local public agencies.

Farm-to-market highway proponents were gradually joined by other activists, some more interested in the development of good trunk roads (main routes). For example, the American Automobile Association was founded in 1902 and by 1907 had joined the National Grange in sponsoring a Good Roads Convention. Eventually, these various prohighway initiatives resulted in a national highway policy codified in the Federal Highway Acts of 1916 and 1921. The 1921 act was based on the concept that a coherent, connected, and comprehensive national highway system could best be achieved through federal coordination. By the 1920s, a Federal Bureau of Public Roads was in place, and limited federal assistance for road construction, mainly in rural areas, was provided.[3] Tolls were prohibited on roads built with federal aid with the exception of high-cost bridges or tunnels, apparently because federal officials thought the reliance on private toll roads in the nineteenth century had been an obstacle to the development of a national road network at that time.[4]

Road construction and finance decisions were still largely left to state and local governments, despite the slowly expanding federal role. The states built the roads financed with federal aid, although they were constructed to federal standards. Federally aided projects were a small portion of the emerging highway system, moreover, since state and local

3. For a fascinating history of the development of federal highway policy and the role of the Bureau of Public Roads see Bruce E. Seely, *Building the American Highway System: Engineers as Policy Makers* (Temple University Press, 1987).

4. Congressional Budget Office, *Toll Financing of U.S. Highways* (Washington, 1985), pp. 7–8.

expenditures still dwarfed federal assistance. During the 1920s an increasing number of states adopted taxes on gasoline as the major source for funding highway construction and maintenance. In many states the new gas taxes were restricted to paying for highway expenses only, a concession to motoring interests who were worried that their tax payments would be diverted to other uses. Through the 1930s the states generally followed the federal lead in using tolls to finance only very special and high-cost links, mainly tunnels and bridges. Since the states were largely responsible for highway decisions, however, they could experiment with alternative methods of financing their road networks, and many different patterns eventually developed.

Public Toll Roads versus Freeways in the 1940s and 1950s

The tolling of roadways as such did not reemerge in any important way until the development of high-performance, limited access highways in the 1930s. Connecticut's Merritt Parkway, opened in 1937, was the first auto-only toll road while the Pennsylvania Turnpike, opened in 1940, was the first high-performance road designed to serve trucks as well as automobiles. Interestingly, 1940 was also the year in which the first Los Angeles freeway (the Arroyo Seco) was built. By the early 1940s successful models thus existed of both tolled and untolled (tax-financed) high-performance highways; the successful toll roads were on the East Coast and the gas-tax-financed freeways were on the West Coast. These patterns of highway development resumed after World War II and by the mid-1950s more than 2,000 miles of toll expressways had been built, mainly in the northeast, while California had completed approximately 300 miles of freeway, chiefly in and around Los Angeles and San Francisco.

The differences in financing strategies were largely explained by differences in the types and densities of traffic. The northeast states were smaller and had more through traffic that might avoid stopping for gasoline and thus helping to finance the roads used. California, by contrast, had less "bridge" traffic because it was located at the end of the continent and was larger. The northeast states also had higher traffic densities than western and southern states, which made toll financing more feasible.

The Interstate Era: 1956 to the 1980s

Development of a truly comprehensive national system of high-performance roads obviously necessitated a compromise or reconciliation of these different regional financial strategies. Such a resolution was achieved with the passage of the Federal-Aid Highway Act of 1956. The act provided funding for the construction of an Interstate and Defense Highway System of 41,000 miles (later increased to 43,000 miles). Ninety percent of the construction cost was to come from the federal grants financed from a federal gas tax of four cents a gallon and increased federal excise taxes on motor vehicles and parts.[5] Following the example of many states, funds from federal fuel and vehicle taxes were deposited in a federal highway trust fund that could only be used for road construction. Tolls were expressly forbidden on the interstate system except for 2,447 miles of toll expressways in operation (or almost completed) at the time the 1956 legislation was enacted.[6]

Several factors led to the decision to finance the interstate system with gas taxes rather than tolls. Tolls were generally opposed by the American Automobile Association and by trucking interests, who argued that the combination of tolls and existing gas taxes would be especially unfair to motorists using tolled facilities (since they would, in effect, be paying twice). Equally critical, however, was the difficulty of financing a transcontinental expressway system through tolls alone. At the request of Congress, the Bureau of Public Roads in 1939 had assessed the prospects for toll financing of a system of six transcontinental roads (three east-west and three north-south). The bureau reported that tolls would be sufficient to cover costs on only 172 miles of the proposed 14,336-mile system and would cover more than 80 percent but less than 100 percent of costs on only another 666 miles. The segments projected to be self-supporting or nearly so were confined largely to the northeast, with one segment each in California and Florida. The bureau went on to recommend instead a 26,700-mile system of untolled access-controlled highways that, to reduce costs, would be built initially with four lanes, two in each direction, only where traffic warranted (mainly in urban areas) and

5. The federal gasoline tax was set at three cents a gallon in 1956 and raised to four cents in 1959.

6. Other exceptions emerged as well, mainly for high-cost bridges and tunnels and for extensions to toll roads that were already part of the interstate system; see Congressional Budget Office, *Toll Financing of U.S. Highways*, pp. 7–10.

with two lanes elsewhere.[7] By the early 1950s the successful operation of toll roads in Pennsylvania, Maine, New York, and Connecticut already suggested that the bureau's forecasts of toll road profitability had been unduly pessimistic. But even the most ardent supporters of toll roads at that time acknowledged that a high-performance transcontinental system could not be financed out of tolls given the light traffic volumes in the vast area between the Mississippi River and California and in much of the southern United States.[8]

Not surprisingly, the establishment of the interstate system discouraged further construction of toll roads. While slightly more than 3,000 miles of toll roads had been opened by 1960, from 1960 to 1980 only about 1,500 more miles of toll road were built (table 10-1). These were mostly extensions of preinterstate toll roads, short urban toll expressways, or a handful of intercity roads in states poorly served by the interstate network (Florida, Kentucky, Oklahoma). Beginning in the late 1960s, tolls were removed on 593 miles, largely under agreements with the federal government to make those roads eligible for federal aid for reconstruction and widening. Consequently, by 1990 only 4,118 miles of toll road were in operation.

By contrast, the construction of untolled expressways accelerated rapidly through the 1960s and the 1970s, only slowing in the 1980s as the interstate system was nearing completion. By 1989 the United States had built 54,145 miles of access-controlled expressways with four or more lanes, and most of this mileage was untolled (table 10-2). The interstate system and its connecting roads accounted for 44,759 miles, of which only 2,695 miles, mainly the core of the pre-1956 toll road network, were tolled. The costs of completing the interstate system systematically increased over the years, as necessitated by inflation, design improvements, and other such influences; however, the basic financing, especially major reliance on gas taxes, remained in place, as did the 90-10 formula for splitting the costs between the federal and state governments.

The Revival of Toll Roads

By the 1980s a point of transition was clearly approaching. The interstate era was coming to an end, and new arrangements for developing the nation's highway system were under active review. In the first

7. Public Roads Administration, *Toll Roads and Free Roads* (Government Printing Office, 1939), especially pp. 5, 155.

8. See, for example, Wilfred Owen and Charles L. Dearing, *Toll Roads and the Problem of Highway Modernization* (Brookings, 1951).

TABLE 10-1. Development of Toll Roads in the United States, Selected Years, 1940–91

Year	*Miles opened*	*Total toll miles*[a]	*Year*	*Miles opened*	*Total toll miles*
1940	166	166	1970	23	3,937
1947	43	209	1971	0	4,002
1950	587	796	1972	70	4,072
1951	167	963	1973	178	4,250
1952	0	963	1974	78	4,328
1953	259	1,222	1975	69	4,397
1954	121	1,343	1976	0	4,397
1955	304	1,647	1977	0	4,368
1956	456	2,103	1978	0	4,368
1957	598	2,701	1979	0	4,368
1958	383	3,084	1980	0	4,368
1959	19	3,103	1981	0	4,368
1960	0	3,103	1982	0	4,368
1961	7	3,110	1983	0	4,368
1962	0	3,110	1984	13	4,306
1963	255	3,365	1985	0	4,177
1964	86	3,451	1986	22	4,199
1965	128	3,579	1987	0	4,059
1966	41	3,620	1988	52	4,111
1967	29	3,649	1989	17	4,118
1968	122	3,771	1990	0	4,118
1969	198	3,913	1991	5	4,133

SOURCE: Unpublished data supplied by the International Bridge, Tunnel and Turnpike Association, Washington.

a. In the late 1960s tolls were removed on 593 miles. The fifty-seven miles deleted in 1969 include forty miles on the Kentucky turnpike and seventeen miles on the Denver-Boulder turnpike. Tolls were removed from these two facilities sometime in the 1960s, although the International Bridge, Tunnel and Turnpike Authority is not sure when.

place, while 90-10 funds could still be used for the widening or rebuilding of the interstate system, Congress had agreed to only minor expansions of the eligible interstate route network. The cost of just maintaining the existing system of highways, high-performance as well as local roads, had outrun the funds available from traditional sources, such as the gas tax, and left few resources for building new capacity. Some federal aid, as well as the states' own taxes, remained available for building or improving noninterstate roads. But in many fast-growth areas, such as the South and West and the outlying suburbs of major metropolitan areas, the growth in traffic was so rapid that available public funds seemed insufficient.

Federal and state gas tax increases covered part of the funding shortfall during the 1980s. The gas tax increases were not enough to offset the effects of the accumulated inflation in highway construction and mainte-

TABLE 10-2. Total Expressway Mileage (Tolled and Untolled), 1956–89

Year	All divided highways with four or more lanes and full access control		
	Rural	*Urban*	*Total*[a]
1956	704	355	1,059
1960	4,687	1,414	6,101
1965	15,238	4,144	19,382
1970	26,727	7,994	34,721
1975	31,889	8,293	40,182
1980	36,608	12,733	49,341
1985	36,836	16,229	53,065
1989	36,900	17,242	54,142

SOURCES: Federal Highway Administration, *Highway Statistics Summary to 1985* (Department of Transportation, 1985), tables SM-211, HM-255; and Federal Highway Administration, *Highway Statistics, 1989* (Department of Transportation, n.d.), tables HM-35, HM-45

a. By 1989 there were 42,064 miles of free highway and 2,695 miles of toll-financed highway for a total of 44,759 miles in the interstate system (highway only).

nance costs, however, and state and local governments began to search for new funding sources. In some rapidly growing areas large private real estate developers began to offer state or local governments right-of-way or cash contributions for new roads to help gain permission for development. A growing number of local governments, particularly in western and southern states, began to formalize these arrangements by establishing schedules of impact fees, or exactions, designed to recover from developers part of the costs of the local road improvements that their projects would require.[9]

Budgetary pressures revived state government interest in toll roads as well. In 1984 Virginia opened the first new toll road in the United States in nearly a decade, the thirteen-mile Dulles Toll Road serving a rapidly growing western suburb of Washington, D.C. Florida opened a new twenty-two-mile toll road in 1986 and began planning major expansions of its toll road system; Texas opened two new toll roads in the Houston suburbs in 1988, and other states began considering toll road projects.

The new interest in toll roads generated pressure to reconsider the federal prohibition on tolling roads built with federal highway aid. The Congressional Budget Office (CBO) seemed to disparage the possibilities of melding federal aid and toll financing, however, in a 1985 report

9. Alan A. Altshuler and Jóse A. Gómez-Ibáñez, *Regulation for Revenue: The Political Economy of Land Use Exactions* (Brookings, 1993).

produced at the request of the Senate Public Works Committee. The CBO acknowledged that toll financing might speed road construction by providing additional capital to supplement tax funds and that toll roads were generally better maintained than tax-financed roads. But, to the CBO, tolls were more costly to collect than gas taxes, and bond financing of toll road debt would force users to pay interest charges (which they did not have to pay under a build-as-you-pay fuel tax system). More telling, the CBO (like the Bureau of Public Roads forty-six years earlier) estimated that toll financing would be feasible on relatively few highways: less than 10 percent of the existing urban interstate mileage carried enough traffic to have been built with tolls, and that percentage would only double if 25 percent federal funding was available. Although the CBO report did not say so, the possibilities for toll financing of additional roads were presumably slimmer since the existing interstate system probably included a disproportionate share of the potentially high-volume roads.[10]

Faced with limited federal aid for noninterstate construction, many states found these arguments unconvincing and in 1987 Congress responded to state pressures by authorizing a pilot program under which up to 35 percent federal aid would be available to construct toll roads in nine interested states (California, Colorado, Delaware, Florida, Georgia, Pennsylvania, South Carolina, Texas, and West Virginia).[11] Only new or substantially improved roads off the interstate system would be eligible as demonstration projects, however, and the states' apportionments of noninterstate federal highway aid would not be increased. Nevertheless, by the end of 1990 three of the states had already started to build their projects, five were in advanced planning or design, and only two had dropped out or were considering doing so.[12]

The prospects for toll roads increased considerably in 1991, when Congress passed the Intermodal Surface Transportation Efficiency Act. The law, popularly known as ISTEA (and pronounced "ice tea"), essentially extended the pilot toll program to all fifty states. Tolls would be permitted on federally aided roads, although still forbidden on the interstate system. Significantly, under ISTEA, states could now also elect to use federal highway aid to assist private toll road projects.

10. Congressional Budget Office, *Toll Financing of U.S. Highways* (1985).

11. Originally there were only seven states; two were added in 1988 and 1989.

12. General Accounting Office, *Highway Financing: Participating States Benefit under Toll Facilities Pilot Program* (Washington, 1990).

Private Toll Road Proposals

The most striking development of the new toll road boom was the revival of the long-dormant idea of private toll roads. Not since the early nineteenth century had private road development of any scale been actively entertained in the United States. The first serious modern proposals for private toll roads emerged in Virginia and California in the late 1980s. The Virginia road proposal, the earliest, was strictly a private initiative, while in California the state organized a competition to award up to four private toll roads as an experiment. In 1991, too, entrepreneurs in Texas filed applications for several private toll road franchises under a nineteenth-century statute (that had never been repealed) authorizing private turnpikes in the state. Also in 1991, the Arizona and Florida legislatures passed laws to allow their state transportation departments to solicit proposals for private transportation; by 1992 Arizona already was in the midst of a competition modeled after California's in which private consortia were competing for up to four franchises.[13]

The Virginia and California Private Proposals

The one Virginia and four California projects, being the earliest, are also the most advanced and thus the most interesting to study.[14] In 1986 a group of private entrepreneurs, later organized as the Toll Road Corporation of Virginia (TRCV), proposed building a fifteen-mile toll road connecting Dulles International Airport with Leesburg, Virginia. The private road would connect with an existing state-owned toll road at Dulles Airport and extend into the rapidly developing western outskirts of the Washington, D.C., metropolitan area. By 1988 the road's backers had convinced the Virginia Assembly to pass the legislation needed to authorize private toll roads in the state. By late 1991 TRCV had received all the state and local government permissions needed to begin construction and was waiting only on final arrangements for construction and long-term financing.

13. See "Arizona Draws Strong Interest in Public-Private Road Projects," *Public Works Financing*, May 1992, pp. 1–7.

14. For a more detailed account of these projects see José A. Gómez-Ibáñez and John R. Meyer, *Private Toll Roads in the United States: The Early Experience of Virginia and California* (Harvard University, Kennedy School of Government, Taubman Center for State and Local Government, December 1991), pp. 21–65.

The California proposals emerged after the state legislature passed Assembly Bill (AB) 680 in 1989 authorizing the California Department of Transportation (Caltrans) to enter into agreements with private companies to build and operate four private transportation facilities (rail lines were eligible as well as toll roads), at least one of which had to be located in northern California and one in southern California. Caltrans organized a competition for private proposals that drew eight entries in 1990.

Caltrans established an elaborate scheme for judging the proposals to avoid complaints of favoritism or political influence. Teams of Caltrans experts ranked the proposals on criteria that included the importance of the transportation need served, the ease of implementation (including environmental or right-of-way acquisition obstacles), the experience of the consortium, the extent to which the project would promote economic development, and the degree to which the project incorporated innovative ideas. The four projects finally selected included three from southern California and one from northern California.[15]

The highest-rated project was from a consortium organized by H. Ross Perot, Jr., the Dallas real estate developer. The Perot group proposed building an eleven-mile toll extension of the State Route 57 (SR-57) freeway from Interstate 5 (I-5) near Anaheim stadium, through central Orange County, to I-405. Caltrans ranked the proposal highly because it would complete a long-needed north-south link in the county's freeway network and would relieve congestion on parallel freeways and local streets. The project was strongly supported by Orange County officials, who alone among California's counties had actively encouraged local AB-680 projects and who regarded SR-57 as one of their top priorities. To mollify adjacent neighborhoods, the road would be built over a channel of the Santa Ana River, a partially concrete-lined river that is usually dry and serves primarily for flood control. Perot named the four-lane, passenger-car-only facility the Santa Ana Viaduct Express (or SAVE) and estimated that it could be built at a cost of $700 million. Tolls would vary by time of day, an innovative feature Caltrans liked, and would range from $5 for each car in the rush hours to $1 for each car in the late evening.

15. For an account of the origins of the California private road program and the selection of the first four projects see Gómez-Ibáñez and Meyer, *Private Toll Roads in the United States*, pp. 67–106; William G. Reinhardt, "AB 680 Special Report: Infrastructure Entrepreneurs Pioneer Private Toll Roads," *Public Works Financing*, October 1990, pp. 1–8, and "AB 680 Update," *Public Works Financing*, March 1992, pp. 14–21.

Second ranked was a proposal by a consortium led by Parsons Brinckerhoff, the giant transportation and environmental planning firm, to extend SR-125 in eastern San Diego County. The SR-125 toll road would run north-south approximately ten miles to the Mexican border and would serve growing residential communities in the eastern end of the county (for example, Bonita and Chula Vista) as well as the increasing truck traffic to and from the Mexican maquiladora plants just the other side of the border. The extension would cost $400 million and would be financed by a combination of toll revenues, contributions of land by local real estate developers, and possibly contributions from local communities along the route. Tolls would be set at approximately twenty cents per vehicle-mile.

Ranked third was another Orange County project, proposed by a consortium headed by CRSS Commercial Group (another large design and engineering firm), for a ten-mile, four-lane toll road completely within the median of the existing SR-91 freeway. The median lanes would relieve congestion on SR-91, which connects rapidly growing residential areas of western Riverside County with the jobs of central Orange County. Orange County had originally planned to build high-occupancy-vehicle (HOV) lanes in the median, and CRSS proposed that the lanes be privately built instead and financed by opening them to toll-paying, single-occupant vehicles (SOVs). HOVs would still be allowed to use the lanes, and those with three or more persons would travel free.[16] CRSS would build the lanes for $88 million, and the costs would be paid entirely from tolls. Tolls would be collected electronically and vary by time of day, with an expected rate of twenty cents for each vehicle-mile during rush hours.

The final project, ranked sixth, was included because it was the highest-ranking project in northern California. A consortium led by the Parsons Corporation (still another large construction engineering company) proposed developing a new eighty-five-mile Midstate toll road from I-680 at Sunol in the south San Francisco Bay area to I-80 near Vacaville. A forty-mile section from Sunol to SR-4 near Antioch would be built first at a cost of $600 million and would provide a high-performance alternative to congested local roads for the developing areas of eastern Contra Costa County and northern Alameda County. The second stage, also costing around $600 million, would be a thirty-five-

16. In its original proposal, CRSS suggested that car pools with two persons would travel free until the road became too congested, at which point they would be tolled. The most recent proposal calls for two-person car pools to be tolled from the outset.

mile extension from Antioch to Vacaville.[17] The second stage would open up areas of largely agricultural central Solano County and provide an alternate route between the Sacramento area and south San Francisco Bay. Caltrans evaluators were attracted to the Midstate because it would pass through a poorly served and developing area of the state, but they were concerned about potential environmental problems, about the difficulties of assembling land and securing support from the seventy-odd communities along the right-of-way, and that the plans for the second stage were highly speculative.

Four other proposals were held in reserve in case any of the four elected projects could not be completed. If Parsons Brinckerhoff's SR-125 project falls through, it will be replaced by a nearly identical, but lower ranked, proposal from a consortium headed by the construction giant Bechtel. If the SR-57 or SR-91 proposals fail, they will be replaced by other southern California proposals: one by a Perini-led consortium to build a $1.3 billion high-speed magnetic levitation rail line between Los Angeles International Airport and the Palmdale Airport (the only non-highway proposal received in the competition) and the second by an Ebasco-led consortium to build a fifteen-mile toll road in Ventura County.[18] The fallback for the Midstate is the only other northern California project submitted: a T.Y. Lin International/Morrison-Knudsen Engineers consortium proposal to replace San Francisco's elevated Embarcadero Freeway, which had been damaged in the October 1989 Loma Prieta earthquake, with a below-grade, toll-free expressway. The $153 million construction cost would be paid back with the profits from building new high-rise office buildings on the right-of-way of the old elevated expressway.

Caltrans negotiated franchise agreements with the four winning projects in 1990. The agreements, all signed by January 1991, set the maximum rates of return that the developers could earn and established zones around the projects in which Caltrans promised not to build competing transportation facilities. Since then, the winning consortia have been immersed in preparing required environmental analyses and trying to win community support; only one (SR-91) has begun construction.[19]

17. Several branches or spurs might also be built, which would add another ten miles to the facility.

18. Perini and Ebasco are large construction firms.

19. For a detailed description of their progress since selection see Gómez-Ibáñez and Meyer, *Private Toll Roads in the United States*, pp. 107–250; "AB 680 Update," *Public Works Financing*, March 1992, pp. 17–21.

These five Virginia and California toll road proposals include a variety of project types. They range in length from ten to eighty-five miles (table 10-3). The Virginia road and two of the California projects (Midstate and SR-125) are development roads, serving areas that are as yet not fully developed but are expected to grow rapidly in the next decade or two. The remaining two California roads (SR-57 and SR-91) are located in a dense urban area and are designed to attract traffic from an already congested public highway system.

The Problem of Financial Viability

The five cases studied here rely primarily on toll revenues to cover costs. They vary, though, in their dependence. Developers of the two roads that would serve already built-up and congested areas (SR-57 and SR-91 in Orange County) claim that tolls will cover all construction and operating costs. Developers of the three roads that would serve newly developing areas (DTRE, SR-125, and Midstate) argue that tolls alone are insufficient and that local government or landowner contributions will be necessary to cover 20 percent to 30 percent of costs.

Political Acceptance of Tolls

Even though tolls are the exception rather than the norm for financing roads in the United States, tolling has not been a source of serious political controversy in most of the five cases studied. The history of the U.S. public toll road movement in the 1940s and 1950s, as well as its revival in the 1980s, suggests that Americans will accept tolling when they perceive that public budgets are constrained and new roads are badly needed. Both Virginia and California adopted highway privatization programs only after state studies revealed a large backlog of needed projects. Both studies recommended gas tax increases, which were adopted, but also forecast unmet highway needs even with these added revenues.

A statewide shortfall alone is often not enough to convince local communities that their road should be among the minority financed from tolls. Two rules seem to apply. Tolls will be accepted only if local residents and officials feel that there is little prospect that their road would be built as a free road and that the process for determining state or local road funding priorities is fair. This fairness doctrine can be under-

TABLE 10-3. Five Private Toll Road Proposals

Name	*Location*	*Miles*	*Estimated cost (millions of dollars)*
Dulles Toll Road Extension	Loudoun County, Virginia	15	300
SR-57 (or Santa Ana Viaduct Express)	Orange County, California	11	700
SR-125	San Diego County, California	10	400
SR-91 median lanes	Orange County, California	10	80
Midstate Toll Road	South Bay to Vacaville, California	85	600 (for first 40 miles)

SOURCE: Developers' proposals as summarized in José A. Gómez-Ibáñez and John R. Meyer, *Private Toll Roads in the United States: The Early Experience of Virginia and California* (Harvard University, Kennedy School of Government, Center for State and Local Government, December 1991). Cost estimates are from original (1990) proposals of the California developers and from the 1991 plan of the Virginia project.

mined, for example, if one community feels that it is being called on to pay tolls when a neighboring community with no apparent greater need is receiving an untolled road.

In two of the five roads studied—Virginia's DTRE and California's SR-57—there have been few objections to tolling. The original Dulles Toll Road was built as a public toll highway a decade earlier,[20] when shortages of state funds made it obvious to local officials that tolls were the only realistic source of finance. Perhaps as a result, the possibility that the extension would be a free road was never seriously raised; the debate focused instead on whether the tolls should be collected by a public agency or a private developer. Similarly, SR-57 had been deleted from California state highway plans in the 1970s largely because the only alignment potentially acceptable to neighboring communities—on a viaduct over the Santa Ana River channel—was widely viewed as prohibitively expensive. Some additional source of revenues, such as tolls, would be needed to offset the high costs if SR-57 was ever to be built.

A brief dispute over tolling did threaten to stop Orange County's SR-91 median lanes. Orange and Riverside Counties originally had planned to build free HOV lanes in the median of SR-91. Riverside County actually had begun construction of its portion at the time that the private consortium proposed building the Orange County segment as a

20. The original Dulles Toll Road is one of the few toll roads built after implementation of the interstate program.

tolled facility. As a consequence, Riverside County argued that HOVs should be allowed free access to the tolled Orange County median lanes, in keeping with the plan the two counties had originally accepted. A guarantee of free access for HOVs threatened the viability of the private proposal in Orange County, however, because HOVs eventually might so seriously congest the toll facility as to make it unattractive to toll-paying SOVs. Eventually Riverside officials compromised and accepted that HOVs with only two persons be tolled because public opinion polls revealed that Riverside residents were more concerned that the lanes be built quickly than whether they were tolled.

Tolls remain an issue for California's Midstate and SR-125. The strongest objections were raised in the case of the Midstate, particularly by California State Senator William Lockyer through whose district the Midstate would run. Senator Lockyer objects to tolls on principle, because in his view they create a two-class system where the well-to-do enjoy high-quality toll roads and those who are less well-off are forced to use inferior free roads. Opposition to Midstate tolls is also fueled to some extent by the long-standing objections of Bay Area residents to tolls on the San Francisco Bay bridges. Nevertheless, Antioch and several neighboring communities on the Midstate alignment support the toll road largely because they perceive state and local resources to be inadequate to finance free high-performance roads needed near their communities.

While not objecting to tolls in principle, some local officials in San Diego have argued that tolling SR-125 is inequitable. The San Diego Association of Governments (SANDAG) had programmed segments of SR-125 north of San Miguel Road for construction as untolled freeways in the 1990s, but construction of the section to the south, which the private consortium now proposes to build as a toll road, was deferred for almost twenty years. SANDAG and Caltrans began environmental planning for the entire length of SR-125, however, which may have given some residents the impression that a freeway south of San Miguel Road was a possibility. Certainly, those residents in the south knew that their compatriots to the north were to receive a free road while their road, if it was to be undertaken soon, would be tolled. A state assemblyman from the area therefore argued that his South County constituents were cheated in the complex countywide negotiations in which SANDAG's priorities were established and has filed a bill that would essentially prohibit construction of the free sections of SR-125 until his objections about the inequity of tolling the southern sections are addressed.

Projects Where Tolls Might Cover Costs

Finding a project where toll revenues will cover most, if not all, construction and operating costs seems to be a far more serious problem than objections to tolling. The basic difficulty is that the United States already has built 54,000 miles of high-performance expressways, including most of those with sufficiently high traffic volumes to be supported from toll revenues. Advocates of new toll roads must therefore search out those remaining unserved opportunities where tolls might cover all or most of costs.

Congestion-relieving roads face different toll and financing problems than roads that service or anticipate development. The congestion-relieving road has the advantage of strong traffic volumes in the early years but often suffers from high construction costs because of high land prices and expensive amenities designed to ameliorate the objections of the built-up communities through which it passes. The congestion-relieving road almost by definition faces competition from the existing expressway network; thus, it can count on heavy traffic only during the peak hours, when the untolled alternatives are normally congested. The least-cost alignments through these built-up areas usually will have already been exploited by public authorities, moreover, leaving private developers only the most difficult and costly alternatives (for example, tunnels, cut-and-cover, or viaducts). By contrast, the development road may enjoy relatively low construction costs per mile, especially where it travels through open country, and it is less likely to suffer from free-road competition. A development road generally suffers the disadvantage, however, of facing a slow and uncertain traffic buildup.

Of the five cases, only the two congestion-relieving roads (SR-57 and SR-91) seemingly rely "entirely" on tolls. However, in actuality, both SR-57 and SR-91 receive nontoll support in that much of their rights-of-way will be provided without charge by state authorities (air rights over the Santa Ana flood channel for SR-57 and land in the median of the existing expressway for SR-91).

The two projects also represent different strategies for providing toll-financed capacity in built-up areas. SR-57 would develop a route that public highway authorities had abandoned as too costly; the construction costs of around $70 million per route-mile (for only two lanes in each direction) are the highest of any of the five projects studied. Given the availability of free alternatives, moreover, SR-57 must earn most of its toll revenues in the peak hours. Even with projected peak period tolls of nearly fifty cents per vehicle-mile, the highest of any of the five projects studied, SR-57 will have to operate near capacity in the rush hours to recover its projected construction costs.

By contrast, the developers of SR-91 identified an underexploited low-cost alternative by building additional highway lanes at grade in an existing freeway median. They project the lowest construction costs per route-mile of any of the five projects (approximately $8 million for two lanes in each direction) and tolls that are more typical of the other projects even in the peak hours (approximately twenty cents per vehicle-mile). The SR-91 project was so obvious that it previously had been planned as a free public project and thus, as noted earlier, the developers became embroiled in a controversy over whether HOVs will be tolled. SR-91 is probably an unusual opportunity; most low-cost new roads in built-up areas have already been built or are already planned as free roads and thus difficult to toll.

A development road typically may find it slightly easier to recoup costs from tolls than a congestion-reliever. However, the DTRE, SR-125, and Midstate cases demonstrate that the task is still formidable. For all three of these development roads, construction costs are projected at around $15 million to $20 million per route-mile (generally for a first stage with two lanes in each direction), a figure higher than SR-91 but much lower than the viaducts of SR-57. Tolls are usually forecast to be between ten cents and twenty cents per vehicle-mile compared with about five cents per mile on existing U.S. toll roads and the almost fifty cents per mile peak toll proposed for the SR-57 project.

Despite lower construction costs, the sponsors of all three development roads still contend that their projects are dependent on contributions from local governments or land developers. The DTRE project seeks developer right-of-way donations worth approximately $60 million compared with development costs (excluding the donations) estimated at close to $300 million as of mid-1991. Projections by SR-125's developers, which may be optimistic, suggest that a donation of approximately $30 million in right-of-way plus contributions of $15.4 million from community facility tax districts might be sufficient to make their $300 million project viable. Midstate developers count on local government contributions of approximately $150 million to $200 million for their $600 million, forty-mile first phase.

Limitations of Landowner or Government Aid

Relying on land developer or local government contributions to make up the shortfall in toll receipts creates its own problems. Developer contributions are most feasible where a handful of large landowners control most of the right-of-way. Strikingly, the DTRE and SR-125 projects, which rely heavily on land donations, are routed through property mostly controlled by a few large landowners while the Midstate, which passes through land held by many owners, relies on local government contributions instead.

Landowner donations often come at a cost of added interchanges, more circuitous routing, or complex three-way negotiations with local governments. Some observers suggest, for example, that the DTRE adopted a more circuitous route to better serve the properties of large landowners. Even so, the DTRE's developers were not able to accommodate the demands of one large landowner who owned a critical part of the right-of-way. He was induced to donate his land only after the DTRE developers brokered a deal with Loudoun County wherein the county granted the landowner certain development permissions he had long wanted. It still took almost three years to complete final agreements for DTRE's right-of-way even though most of the needed land was owned by a few large developers who favored the project.

The SR-125 experience has been similar to that in Virginia. Two major landowners made general statements about their willingness to support the project during the competition among AB-680 proposals but avoided making specific commitments. When the SR-125 consortium began negotiating in earnest after the competition, the real estate market in San Diego had softened and the landowners were less forthcoming. They reportedly insisted that their contributions be credited against the development impact fees that they would normally be obliged to make to the city of Chula Vista, a position that the city was initially unwilling to accept.

Like local landowners, local governments often have a stake in a proposed road and thus may be induced to contribute if they are convinced that contributions are necessary. Some of the local government authorities that the Midstate consortium had hoped would contribute have proven uncooperative, however, largely because they are not convinced of the need for the road. Governments may also condition their contributions on design changes or other modifications that add to costs. In the case of SR-125, Chula Vista, after a long dispute, agreed to allow some development impact fees to be used by the private project, but these would be counted as an equity investment by the city rather than granted for free.[21]

None of the private toll roads studied here rely on federal aid since federal law prohibited charging tolls on roads built with federal assistance, except under special and limited circumstances, at the time the projects were developed. With the 1991 changes in federal law, however, states can now use federal highway funds to pay 50 percent of the cost of building a toll road, public or private, provided that the facility is new or substantially improved and not part of the interstate system. A 50 percent subsidy obviously increases, perhaps significantly, the number of situations where toll roads are financially viable. Such federal aid could eliminate the need for landowner or local government assistance required by the three development roads studied, for example, or reduce the required tolls on SR-57 to more manageable levels.

Congress left the decision to use federal aid for toll roads to the states, however, and provided no special toll road funds above and beyond each state's normal apportionment of federal highway aid. The use of federal aid for a toll road, public or private, therefore will come at the expense of its use for other road projects. Proponents of toll roads might argue that their projects would make federal aid go farther, but this argument would be compelling only in states that did not have enough state or local tax receipts to match federal aid.

Federal aid may also impose some additional costs on toll road developers. State officials are likely to insist on certain modifications or guarantees in return for the federal aid, much as local officials and landowners do.[22] The general lesson seems to be that relying on nontoll revenue sources usually complicates projects, which are often difficult enough already, by adding new sources of potential controversy or delay.

21. "AB 680 Update," pp. 19–20.

22. Recipients would have to comply with federal regulations governing environmental reviews, contracting, and other matters in addition to the comparable state or local regulations that already apply.

Are Private Roads Cheaper, Faster, or More Innovative?

Our five U.S. case studies cannot provide definitive evidence on whether private road construction or operation is cheaper than public, largely because none of the five has been built or placed in service.

Construction and Operating Costs

The only explicit comparison of public and private costs was made in Virginia, where the State Corporation Commission (SCC) was required by law to compare public and private costs and timeliness before awarding a franchise to a private road developer.[23] The SCC staff comparison, summarized in table 10-4, claimed that a toll road built by the Virginia Department of Transportation (VDOT) would cost motorists only $0.9 billion over forty years, while TRCV's private road proposal would cost $3.5 billion. The VDOT road would cost more to build ($211 million versus $153 million) largely because the SCC assumed that landowners would donate right-of-way to the private road but not the public road. In the long run, however, the SCC estimated that the private proposal would cost users more because it was to be financed with a sale-leaseback scheme with an effective annual interest rate of 10 percent; the VDOT road, by contrast, was assumed to be financed by a combination of 7 percent tax-free state bonds supplemented by $22 million in surpluses from the existing Dulles Toll Road (to cover debt payments in early years, before traffic built up). TRCV would also have to pay federal and state income taxes, local property taxes, and dividends to its equity investors. As a consequence, the SCC projected that a VDOT road would require an average toll of only $1.00 for each vehicle over the forty-year life of the highway, whereas TRCV projected that its initial toll of $1.50 would rise in stages to $3.25 by the year 2010.

Such simple cost comparisons are misleading, however, in that many of the reported differences in VDOT and TRCV costs do not represent fundamental savings to society but rather are transfers from one part of society to another. A private road would be less costly to society as a whole only if it required fewer physical resources, services, or amenities to build or operate than a comparable public road. The private road might be less costly, for example, if it required less right-of-way, concrete,

23. Gómez-Ibáñez and Meyer, *Private Toll Roads in the United States*, pp. 31–54.

TABLE 10-4. The Virginia SCC's Comparison of Public and Private DTRE Projects
Millions of dollars unless otherwise noted

Item	Public	Private (TRCV proposal)
Project costs through construction		
Direct construction		
Construction	154.3	145.4
Right-of-way	49.5	0.0
Engineering and design	7.0	6.0
VDOT charges to TRCV	0.0	1.8
Subtotal	210.8	153.2
Other		
Development/administration	2.5	13.6
Financing fees, construction interest	22.4	30.4
Taxes during construction	0.0	1.6
Subtotal	24.9	45.6
Grand total	235.7	198.8
Financing		
Revenue bonds (7%, thirty years)	218.0	0.0
DTR surpluses	22.3	0.0
Sale to owner (for lease back)	0.0	198.8
Equity	0.0	30.0
Tolls	$1.00 constant	$1.50 initially, rising to $3.25 by 2010
Direct costs of service (forty years undiscounted)		
Debt service	456.5	916.0
Operations and maintenance	438.1	640.0
Federal and state taxes	0.0	689.0
Local taxes	0.0	96.0
Dividends to stockholders	0.0	1,127.0
Total	894.6	3,468.0

SOURCE: Commonwealth of Virginia, State Corporation Commission, "Staff Report: Application of the Toll Road Corporation of Virginia, Richmond Case No. PUA900013," April 17, 1990, pt. A, tables 5, 6, p. 45.

or labor to build because of more efficient design. The leftover land, labor, and concrete could then be used for other projects, such as building another road.

Most of the claimed savings for VDOT or TRCV are not of this fundamental type, since VDOT and TRCV's designs, alignments, and direct construction costs seem similar. In the SCC analysis, VDOT exhibits savings largely because it is tax exempt and uses various financing devices to shift risk from road investors and users to taxpayers. VDOT's tax-exempt debt pays only 7 percent while TRCV's taxable lease pays 10 percent, for example, but this savings does not represent any inherent difference in the riskiness of the two projects. Rather, VDOT's tax-exempt debt just shifts part of the costs from DTRE investors and motorists to federal and state taxpayers (in the form of lower tax revenues).

Similarly, VDOT's use of surpluses from the original toll road to cover early-year losses on the extension and bonds backed by the full faith and credit of the state, not just toll revenues, only shifts underlying risks to state taxpayers rather than reducing them. VDOT in effect would be asking taxpayers to make a risky equity investment in the project (by pledging toll road surpluses and the full faith and credit of the state to cover early-year losses) but not compensating them for the risks they were assuming. One might argue that the state taxpayers should receive the same 21 percent return on their investment of toll surpluses that TRCV estimates its equity investors will require, since the underlying risks are much the same. In that case, calculations suggest that VDOT would have to charge a toll similar to that being proposed by TRCV to recover its debt and "equity" costs. Indeed, VDOT tolls might be slightly higher than TRCV tolls if, in addition, VDOT used bonds backed by toll revenues alone and not by the full faith and credit of the state. By the same token, TRCV's savings from the donated right-of-way are also transfers (from landowners to road users).

The one real cost advantage claimed in the TRCV-VDOT debate was TRCV's argument that it could open the road sooner by starting construction earlier and building it faster. If true, this advantage would represent a real savings to society. All else being equal, starting sooner and building faster reduces the amount of capital that must be tied up during construction and before the fruits of the investment are enjoyed. TRCV's early experiences are somewhat discouraging, however, since obtaining the necessary public approvals and the search for financing has taken five years. Part of the delay is because DTRE is a pioneer private toll road project in the United States; few precedents exist to guide developers or public officials. Nevertheless, some tasks—such as negotiating with landowners for contributions or with Loudoun County for approvals—took far longer than TRCV expected. Moreover, VDOT had nearly completed the planning and environmental approvals for its competing public alternative when it abandoned the effort to TRCV in 1990, which suggests that it might have been able to begin, but not necessarily complete, construction on about the same schedule as TRCV.

The main lesson of the Virginia comparison is that transfers may be more important than real savings in cost debates about road privatization. In the Virginia comparison, which is "cheaper," the public or private route, depends largely on who you are. A prospective motorist on the DTRE is probably better off with the public rather than the private proposal, since the public proposal shifts costs to state and federal taxpayers by using tax-exempt financing and by exploiting a hidden and uncompensated equity investment to cover early-year losses. From the perspective of the state or nation as a whole, however, the public and private costs appear roughly comparable.

Overcoming Environmental and Land Assembly Problems

All new roads in the United States, public or private, face potential obstacles focusing on their environmental impacts, the growth they might induce, and the land takings they require. Private provision of roads might be advantageous if private developers could resolve these problems more easily or responsibly than a public agency.

Virginia and California decided that private toll roads would have to comply with the same environmental impact laws as a public road. Virginia offered local governments veto power over a private road as well, by stating that a state certificate would not be granted over the objection of an affected local government. Neither Virginia nor California gave the private developers the power to acquire right-of-way against a landowner's will through the power of eminent domain, although the Virginia legislature allowed local governments to exercise eminent domain on behalf of a private road and the California legislature allowed Caltrans to do so. (Caltrans has indicated that it would only do so as a last resort.) As of 1993, only Virginia's DTRE and California's SR-91 had secured all needed environmental and local government permissions, although the other three California roads were well into the process.

The private backers of these five roads clearly were sensitive, when making their proposals, to potential environmental, growth management, and land assembly objections. Nevertheless these projects face potentially significant problems. For example, both of the congestion-relieving roads, SR-91 and SR-57, face difficult air quality issues because of the considerable air pollution in the Los Angeles area. SR-91 has surmounted these objections reasonably easily and plausibly by arguing that free-flowing traffic, which the SR-57 and SR-91 projects should abet, is less polluting.[24] For SR-57, a more serious problem is likely to be the noise

24. "AB 680 Update," p. 17.

and unsightliness created by its elevated viaducts: the apparently agreed-on alignment may prove less acceptable when people have a closer look at preliminary models and renditions.

The three development roads have raised surprisingly few environmental objections, although in the two California cases this situation may simply mean that their alignments and designs are not fully developed. Virginia's DTRE, a quintessential development road, faced objections to its crossings through wetlands and the design of a major bridge over a river along its alignment, although these problems were eventually solved. San Diego's development road, SR-125, also looks manageable from an environmental perspective. One problem, disturbance of the nesting areas of some endangered bird species, can probably be met by relocation of the right-of-way, for example, since the road passes through largely open and undeveloped country so that right-of-way options are not too restricted.[25]

The Midstate highway appears to face the most substantial environmental problems of the roads studied. Few environmental studies have been done on this alignment, in large part because the road is by far the most ambitious and original of any of those sanctioned in the California competition. The Greenbelt Alliance and the Sierra Club have joined forces to oppose the road, but they seem inspired more by general antigrowth sentiments than by environmental issues relating to the Midstate highway itself. Again, much like SR-125, the environmental issues that arise may be susceptible to fairly simple solutions by changing the alignment, since the Midstate is also being largely built through relatively open country.

The five cases suggest that private roads have advantages and disadvantages in overcoming potential environmental, growth management, and land assembly objections. The primary advantages are the private sector's great sensitivity to the cost of delays and their greater flexibility in finding compromise solutions. The most dramatic example is DTRE's successful efforts to overcome wetland and river crossing problems. Ralph Stanley, the chief executive officer of TRCV, approached environmental groups in a cooperative and generous spirit, which he called "aggressive compliance." He offered to replace wetlands on a two-for-one basis, instead of the 1.5-for-one that usually characterizes land

25. The community of Chula Vista filed an environmental suit, but this action may have been more of a bargaining ploy than a reflection of serious environmental concerns; as already noted, Chula Vista, for several reasons, apparently feels somewhat shortchanged owing to previous road funding negotiations in San Diego County.

swaps for conservation purposes, and to build a bridge with one long span and no piers to alleviate concerns about the impact on a river's scenic beauty. Since time was money for TRCV, quick and generous compromises were much more sensible than protracted litigation. Similarly, a private firm may be able to avoid eminent domain where a public agency could not. California's right-of-way acquisition laws limit the state to paying fair market value for a parcel, for example; private firms are not so bound and may pay more to avoid protracted eminent domain proceedings.

Private developers also face disadvantages. In the first place, their greater sensitivity to delays is a handicap as well as an advantage. The backers of each of these roads estimate that preparing the analyses and designs necessary to secure environmental and other permits will cost them several years and $10 million or more in each case. DTRE's backers spent three years and about $20 million to secure permits for a road with comparatively little opposition. Raising venture capital for this effort is extraordinarily difficult since the risks are perceived to be high. DTRE's backers were lucky to find a wealthy individual with faith in their project, while the other projects have had to rely primarily on contributions of "sweat equity" from the planning, design, and financial firms in their consortia. Thus it is not surprising that private backers are more ready to compromise than litigate, since the prospect of serious controversies or delays can easily become a deal breaker by frightening away investors at the development stage.

Private developers may also generate more suspicions than a public agency might, especially when growth management or land assembly is an issue. In matters of growth management, for example, a private toll road may raise the specter of collusion with real estate developers. This suspicion is reinforced when, as in two of the three development roads studied, the project backers have sought contributions from developers who own large adjacent parcels of land. Private toll road developers have tried to mitigate these potentially adverse perceptions in various ways. For DTRE, TRCV tried to sidestep the problem by saying that it would not let landowners become equity investors, although they could donate right-of-way. The Midstate developers constantly reminded local governments that they did not intend to undertake any land development and could not do it anyway without local planning permissions. Nevertheless, suspicions almost inevitably arise, at least among those inclined to anti-growth sentiments, whenever private road developers and landowners come together.

Finally, the perception of flexibility may sometimes work to the disadvantage of private road developers by raising issues that might not have occurred with a public highway authority. For example, that private road developers are not constrained to pay fair market value by state compensation statutes may also encourage landowners to hold out for much more. And sometimes local communities along the route may insist on assistance for their feeder road systems that private developers doubt would ever be asked of a state highway agency.

Innovation

The Virginia and California projects suggest that privatization does enhance innovation in the selection, design, and operation of toll roads. For example, three of the five proposals studied had been rejected, neglected, or overlooked by public highway authorities. The extension of SR-57 had been abandoned by Caltrans as too expensive in the 1970s, the southern segment of SR-125 was on state and county plans but was not to have been built for at least ten or twenty years, while the Midstate alignment had not even occurred to public officials.

The designs and operating plans of the private projects are innovative as well. The sponsors of SR-57 have proposed an auto-only road, for example, to reduce the cost of building the viaduct and thus help make this expensive project financially viable. SR-91 is the first project in the United States proposing that toll-paying SOVs use special HOV lanes, a concept that potentially provides a means not only of financing HOV lanes but of exploiting the often underused capacity of these facilities as well. Another idea being pioneered by these private projects is varying toll rates with time-of-day or congestion. Peak-hour or congestion pricing is a key element in the plans for the two congestion-relieving roads, SR-57 and SR-91, and is being considered by the three developmental roads as well. It is striking that no U.S. public toll road has ever adopted time-of-day or congestion pricing, despite the advantages often cited by economists and widespread use of such schemes in airlines, telecommunications, power, or other industries whose service demands vary seasonally or hourly.

The potential importance of innovation has implications for the design of privatization programs. Some have suggested, for example, that state governments should assume the responsibility for conducting environmental studies and winning local permissions, since raising the venture capital needed poses difficult problems for private road promoters. The approved projects would then be turned over to private interests for consummation. This action clearly would make private investment more available, but it would also reduce the opportunities for innovative projects. Virginia's and California's strategies of allowing the private sector to identify and propose projects exposes the firms to more developmental risks but also gives them the flexibility and incentive to develop new ideas.

The Problems of Monopoly and Regulation

Surprisingly little controversy has arisen over the potential problems of monopoly or market competition that might arise with private toll roads. California chose to grant the private toll roads exclusive franchises, delimiting territories within which the state promised not to build competing facilities and to use its best efforts to persuade local governments from doing so as well. Virginia's DTRE did not request similar protections, probably because it thought improvements to parallel facilities were unlikely on budgetary and environmental grounds.

Both California and Virginia decided to regulate profitability directly by regulating the allowable rate of return on investments instead of, or in addition to, regulating toll rates. Virginia adopted conventional rate-of-return regulation by the state public utility commission (PUC) that also regulates electricity, telephone, gas, and other utility rates.[26] In this system, requests for tariff increases are granted only if a PUC investigation determines that without them the regulated company's return on investment would be lower than that being earned by other private and competitive industries facing comparable risks. California selected an innovative one-time contractual approach, setting the ceiling for the rates of return allowed over the life of the project in each franchise agreement. California's private toll road operators are free to set their own toll rates so long as they stay within the ceiling returns set out in their franchise agreement.

So far, both state officials and the private road developers profess to be satisfied with these arrangements. The only complaints have come from opponents to a few of the California projects, who argue that the maximum rates of return negotiated in the contracts are excessive. One of the California projects (SR-91) completed financing and began construction in 1993, which suggests that California's innovative approach is acceptable to private investors.

26. Virginia's PUC is the State Corporation Commission mentioned earlier.

In regulating rate of return directly rather than indirectly through tolls, Virginia and California differ sharply from Spain, France, and most developing countries that have private toll roads. Since the rate of return is the fundamental concern of both investors and users, this direct approach probably has a lot of advantages.

Moreover, California's one-time contractual approach seems to offer important advantages over Virginia's PUC model. California's system does not require cumbersome and potentially controversial rate-of-return investigations after every request for a toll increase; all the state has to do is audit the accounts to insure the target rate of return is not being exceeded. California's contractual approach also ostensibly fixes rates of return for the life of the project, thus protecting investors from changes in the political climate that might affect a PUC.[27] It seems unlikely, however, that a contract signed in the early 1990s can anticipate and cover every contingency through the 2030s (when most of these franchises will expire). Thus some exposure to the vagaries of the political climate may be unavoidable.

The California calculations of ceiling caps on earnings may also become controversial. California agreed to fix ceiling rates of return on combined debt and equity of 17 percent to 21.5 percent over the thirty-five-year lives of each project (each of the four projects has a different fixed cap). By contrast, Virginia's PUC set a ceiling return of only 14 percent and only for the first six years. Whether or not California's higher rates are necessary to attract capital is unclear. On the one hand, it might be argued that the Virginia PUC, unaccustomed to dealing with infant industries, may not recognize the potential risks of pioneering a private toll road. On the other hand, Virginia's DTRE project is arguably less risky than some of the California projects, particularly Midstate and SR-57. Virginia's 14 percent rate was also the rate that DTRE's backers had requested and presumably thought adequate (although they could prove wrong).[28]

27. Virginia's regulators agreed to protect investors to some degree from political developments, however, by establishing an account in which the unrealized earnings from the early and riskiest years of the project accumulate as liabilities for later repayment out of earnings.

28. Moreover, the California ceiling cap seems overly sensitive to the scenarios that the state's financial consultants assumed in their calculation. For example, in doing the calculations undue weight may have been placed on the hazard of permits being denied at the last minute. In the real world such adverse possibilities should be evident well ahead of time so that private sponsors can pull out before they lose all potential development expenses; Gómez-Ibáñez and Meyer, *Private Toll Roads in the United States* pp. 97–100, 201–02.

Moreover, by capping the rates of return, California may limit the incentives for the concessionaire to control costs or not to abandon the concession once they have reached their ceiling return. Although perhaps not intended to do so, California has included "incentive returns" that may alleviate this problem. All four private roads can earn up to 6 percentage points above their allowed ceiling rates of return if they meet certain public objectives, including increasing average vehicle occupancy, reducing toll road operating costs, or reducing accident rates. One issue is whether incentives for such purposes are necessary or too generous.[29]

California might have solved the abandonment problem more directly, moreover, by allowing profit sharing after a certain rate of return rather than capping the return. Such an approach was used in a concession for a private bridge in Puerto Rico in 1991, for example. In Puerto Rico the franchise agreement states that if returns exceed 18 percent, 60 percent of the excess goes to the state; if they exceed 22 percent, 85 percent of the excess goes to the state. This scheme effectively limits returns above 18 percent but not so completely as to eliminate the incentives for a concessionaire to continue to operate the project.[30]

Prospects and Opportunities

What do these five case studies suggest about the opportunities for private sector toll roads elsewhere, especially in the other forty-eight states? To start, they indicate that development roads may have a slightly better chance of being privately developed than congestion relievers, although much depends on local or site-specific considerations. Financing development roads with tolls is probably likely to be politically acceptable and can be accomplished at costs that are not so exorbitant that they cannot be recovered from tolls. The congestion relievers, by contrast, are almost by definition very high-cost facilities (with SR-91 an exception that proves the rule). Congestion relievers often will need to tap into other sources of financing that will greatly complicate the task of estab-

29. The case for vehicle occupancy incentives seems the most plausible, since increasing occupancy is arguably more in the interests of the general public than the private operator. By contrast, operating cost reductions and safety seem to be in the operator's interest and largely beyond the operator's control. The incentives may be too generous in that operating expenses are a comparatively small share of total roadway costs, for example, and the vehicle occupancy targets may be too easily met.

30. For a description of the Puerto Rico system see William G. Reinhardt, "Case Study: Puerto Rico Toll Bridge Concession, Does it Count as a Private Road?" *Public Works Financing*, April 1992, pp. 14–16.

lishing the political alliances and coalitions needed for implementation. Furthermore, the congestion relievers, since they go through already built-up areas, will have fewer opportunities than the development roads to tap into the most obvious source of outside financing, that of donations from private developers holding large, nearby, vacant land holdings.

A far more robust generalization, however, is that private toll roads probably will not be built in very large numbers in the United States given the constraints of financial feasibility and political acceptance. The most important limitation on the building of private toll roads is the sheer size of the existing U.S. road system, which means that many of the potential opportunities for profitable toll roads have already been preempted, enhancing the difficulties of attracting landowner contributions or government aid to supplement tolls. Tolling may also not be politically acceptable in some situations, and the high costs of delays for private investors will make them avoid roads with significant opposition because of environmental or siting problems.

Nevertheless, private toll roads may make a contribution in the United States by stimulating innovation and by serving as a benchmark against which the performance of public highway authorities can be measured and stimulated. The principal advantages are not likely to be in lower-cost construction or operation or in bringing some roads on stream faster than the public sector could, although these beneficial results may sometimes occur. The main contribution of private toll roads is more likely to be the willingness of their builders to be innovative and to explore new technologies and techniques. These potential contributions, while no solution to all highway problems, are not trivial.

Chapter 11. Standardization, Regulation, and Other Trade-offs

THE PRIVATIZATION OF HIGH-PERFORMANCE highways raises many of the same public policy issues as privatization of urban transit buses: competition, financial viability, efficiency, and equity. In the case of highways, however, four issues assume added importance or raise other complications: the pressure for standardized service or networks, the problem of regulation, the roles of innovation and efficiency, and the problem of facility siting.

The Problem of the Standardized Network

Pressures for a standardized level of service, including equalized prices or tolls, over a broad geographic area affect the privatization of bus transit, mainly in concerns about whether network coverage will be maintained after privatization. Such considerations, however, are far more prominent in highway privatizations. For example, the perceived need for a national network of high-performance highways designed to a common standard helped shape the French autoroute companies and greatly influenced the Spanish and U.S. decisions to switch from toll to tax financing of highways. Network standardization is emerging as an issue in Mexico's private highway program as well and is likely to become important in other developing countries as they seek to expand their highway networks outward from high-density urban areas to remote lower-density locations.

Pressures for physically standardized and uniformly priced networks are common in other forms of infrastructure besides highways, moreover, including pipelines, railroads, airlines, telecommunications, public health, and water supply and sewage systems. For many of these activities, the ostensible advantages of a ubiquitous standardized network are often used to bolster the argument for government involvement or even government ownership. Accordingly, any privatization initiative almost always becomes inextricably linked with these network considerations.

The Merits of Standardization

Fundamentally, the ubiquitous standardized network can be seen as a way of politically or socially integrating a nation, especially where there are substantial regional income differences or important ethnic or national minorities. Standardized networks are often seen as helpful, for example, in achieving a more uniform pattern of regional economic development.

Efficiency arguments can be made too. A nationwide homogenous standardized highway or rail network permits standardization of fleet equipment and services. Obviously, disruption is more likely if truck operations have different allowable axle weights or trailer dimensions on different parts of a highway system. Similarly, changes in gauge dampen the productivity of rail networks. In telecommunications, a homogenous network facilitates efficient standardization of equipment, codes, and interconnections.

Favorable externalities are often cited as benefits of ubiquitous standardized networks. A ubiquitous telecommunications system may make it easier to insure the public safety, health, and welfare of some of the more dependent groups in society. With telephones everywhere, crime should be more promptly reported, for example, and an accident on a country road is more likely to elicit a prompt emergency response if the neighboring farmhouse is equipped with a phone. Furthermore, many people argue that ubiquity, standardization, and universality make almost any network or system more valuable to all, by creating more options that have value even if exercised only infrequently.

Efficiency arguments can also be made against the ubiquitous standardized network. Obviously, standardization eliminates flexible adaptation of capital to needs. For example, a major concern during the congressional debates leading up to the authorization of the Interstate and Defense Highway Program was that four-lane, limited-access divided highways were probably not needed, at least immediately, in much of the western United States. Similarly, a radio phone might be a more cost-effective way of meeting telecommunications needs in isolated areas than the conventional hard-wired copper line. In essence, the ubiquitous standardized network can create artificial technological or capital constraints that otherwise might be avoided.

Significantly, though, strong political support for the ubiquitous standardized network arises even when the economics run strongly to the contrary. The decision to build the U.S. interstate system almost everywhere to minimal standards of four lanes, divided and with limited access, is perhaps the most obvious example. The French experiments with cross subsidies, both within and between highway concessionaires, to develop high-capacity, high-performance highways in very remote areas further indicates the political attractiveness of the ubiquitous standardized network.

The political attractions of the ubiquitous standardized network are also illustrated by the principle of exclusion: it is difficult to identify many instances where the requirement, once adopted, is later relaxed. One of the few such instances was U.S. local air carrier service. Conventional jet services to most small communities were replaced with more frequent jet-prop plane services after U.S. airlines were deregulated in 1978. Similarly, since the breakup of American Telephone and Telegraph's monopoly and the relaxation of telephone regulation in the United States, greater flexibility has been permitted in technological specifications for telephone services to more remote locations. But the exceptions are few and far between. The political support for ubiquitous and standardized public service infrastructure networks is a remarkably consistent and international phenomenon, often overriding even adverse economics.

Privatization versus Standardized Networks

Ubiquitous standardized networks, popular as they may be, are often difficult to reconcile with privatization. Private enterprise, without subsidies or other subventions, normally will only build profitable segments of a network, that is, those that pay their way from fares or tolls. In practice, this rule means that private enterprises prefer those links where demand is already well established and relatively dense (since, with standardization, the costs of building a link are not likely to vary that much between low- and high-demand situations). Private vendors, then, tend to skim the cream by selecting only the most economically advantageous segments for development. Various subventions are usually needed to build parts of the system that are economically less attractive; for example, cross subsidies or donations by other interested parties, such as real estate developers, who potentially can profit from development of the network. Unfortunately, subventions, at least from the private sector, are likely to be more available where they are least needed (that is, where network profit potentials are highest). Consequently, socially or politically important segments of the ubiquitous standardized network are likely to remain undone if undiluted private enterprise is the exclusive agency for development.

Several solutions for filling in lost or orphan segments are possible. One has already been implied: degrading or accepting lower levels of service at more remote, isolated, low-income, or other disadvantaged locales. As already noted, that solution is not likely to be politically popular.

A second possible solution, and probably the most commonly adopted, is simply to rely to some extent on government ownership and finance of the needed facilities. At one extreme, the entire network can be "nationalized," using the deep pockets of the taxpayer as the ultimate source of funding. Nationalization has been a widely accepted solution: northern Europe used it for its roads and highways, the United States followed such an approach for its roads and highways between roughly 1850 and 1940 and then again after 1956 when the interstate highway system was authorized; the world outside of North America, at least until recently, has almost universally used government-owned operations to provide railroad, airline, and telecommunication networks; and even in North America, transit and postal services have usually been provided by a public agency.

With government ownership, a more limited or segmented approach also can be pursued, that of providing only the missing links not offered by private enterprise. In fact, that apparently happened in the United States during the first half of the nineteenth century (with public roads filling in where private toll roads were not built). The Spanish pursued similar policies during the 1980s by supplementing the private and tolled autopistas with public and tax-financed autovías. The government-owned Canadian National Railroad is still another example, since its historic role was largely to fill in or supplement the rail service provided by the private Canadian Pacific Railroad. Domestic airline service in some developing countries, provided at fares and demand densities that are uneconomic for private provision, have often been continued in the public sector even as the same countries have privatized their international air services.

Still another possible solution to the inherent conflict between privatization and achievement of the ubiquitous standardized network is to adopt cross subsidies. As noted when discussing France's toll road experiences in chapter 8, cross subsidies can be internal (that is, within an enterprise) or between enterprises. The internal or within-an-enterprise approach has a long and hoary history in transportation. The Interstate

Commerce Commission, for example, developed a highly elaborate set of proposals for merging unprofitable railroads with the profitable in the 1920s, although these plans eventually foundered on the understandable reluctance of the profitable railroads to accept such mergers. France's highway authorities were more successful in promoting such mergers, perhaps because most of the autoroute companies were publicly owned. Not surprisingly, therefore, transport franchising of private vendors by governments probably means many instances of attempting to balance a portfolio between losing and profitable operations.

Cross subsidization between different enterprises is less common and more difficult than within an enterprise, usually requiring some third-party umbrella organization. When taxes are the basic source of financing, government itself, of course, commonly performs this function. When relying on toll financing, however, some special agency is usually required. Thus the French created a toll road association for this purpose. Similarly, in U.S. telecommunications, AT&T presided for several decades over a so-called separations procedure that divided up the substantial profits of long-distance services to compensate local telephone companies for the lesser profitability of local services. For highways, transfers sometimes can be effectuated from enterprises outside the industry to enterprises within, for example, from real estate developers to highway developers; in fact, as noted in chapter 10, that pattern of cross subsidization is being pursued in several private toll road proposals in the United States.

Cross subsidization, especially when pursued between or within enterprises in the same basic activity, usually has implications for regulatory policy. Cross subsidization normally requires different markups over cost for different activities; this, in turn, makes the high markup activities far more attractive entry possibilities to alternative outside suppliers. With the passage of time, moreover, buyers have strong incentives to find alternative suppliers or technologies that avoid or erode the high markups; such erosion clearly imperils the cross subsidization. Consequently, schemes for cross subsidization tend to be accompanied, either before and almost certainly after the fact, by regulatory arrangements aimed at stabilizing the sources of cross subsidy. These stabilization schemes often create problems, including impediments to technological change or innovation.

Still another way of resolving the potential conflict between privatization and network development is to have the government directly subsidize private suppliers of the missing links. Such subsidies can be provided in several different ways. The subsidies might be done by formula as in the Civil Aeronautics Board's promotion of local U.S. air service before airline deregulation. Or competitive bid or auction may work. The British used this approach with certain noncommercial bus services after the 1986 reforms, and so did the Mexicans in their most recent rounds of private highway concessions. Finally, subsidies might be set in the old-fashioned political way as part of a pork barrel negotiation between interested parties; apparently, decisions about Amtrak passenger rail service in the United States have sometimes been made on this basis.

Which of these solutions works best? Obviously, the answer will greatly depend on the goals and circumstances. One guiding principle, though, does seem applicable: solutions should be imbedded in reasonably consistent policy packages. For example, privatization and reliance on market mechanisms, such as auctions, for distributing any subsidies is probably most acceptable and consistent when innovation is an important public policy concern. By contrast, government ownership or nationalization is more easily rationalized, if not justified, when externality concerns weigh heavily. Mixed systems, say, combining cross subsidy with some private ownership, can be made to work, as the French highway experience well illustrates, but these hybrid systems are often difficult to implement administratively and usually cannot be effected without considerable loss of economic benefits.

Competitive and Regulatory Complications

Government regulation of prices or allowable returns on investment is a far more important problem when privatizing highways than urban buses, if only because competition is so much harder to maintain for highways. Bus transit is characterized by few scale economies; it is therefore easy to envision entry, or the threat of entry, so that competitive, or at least contestable, markets seem plausible. Obviously, by comparison, entry into the activity of supplying highways is likely to be restricted. It is difficult to imagine more than two or three sensible alternative highway routes going between most points. Accordingly, a truly competitive or even contestable market in intercity highway services does not seem likely.

Virtually every country that allows privatized roads requires that an alternative free route be available, but this condition is seldom a satisfactory solution to the problem of competition. On the one hand, too much free road competition can result in inefficient allocations of traffic (where the free road is too heavily congested and the toll road is underutilized) and can undermine the financial viability of a private toll road. On the other hand, too little competition can open up the possibility of the private developer earning excessive returns (that is, more than needed to keep the necessary capital committed). Most governments therefore usually retain some regulatory oversight of private highways to prevent monopoly abuse even when a free alternative is in place.

Regulation is not only more necessary but also more difficult with highways than buses. Highways are so capital intensive and the investments so long-lived that investors need assurances that the regulatory authorities will be fair and stable over the long period needed to earn an adequate return. Highway regulation is further complicated because balancing congestion on parallel free roads and establishing a ubiquitous standardized network, with standardization commonly including not only physical characteristics but a uniform toll rate as well, are often important considerations. When compounded by political constraints and influences, the policy complications may approach the insurmountable.

The highway cases reviewed in chapters 8, 9, and 10 illustrate several approaches to these regulatory problems, some more satisfactory than others. Probably the most common is to specify the allowable toll rates in the franchise contract, usually allowing some adjustment with general inflation or other specific cost escalations. The long lives of highway investments make it both difficult and critical to reach agreement on the adjustment formula, however. The French and Indonesian governments were reluctant to commit themselves to a formula, for example, for fear it might eventually prove too generous. Without a commitment, however, private investors have been understandably discouraged from making long-term commitments themselves. At the other extreme, the Spanish have committed themselves to a specific industrywide formula for toll increases, but, barring any explicit assessments of the rates of return needed and earned, this formula is probably too generous for some roads and too stingy for others. Essentially, fixing tariffs and escalation formulas in the franchise agreement is a fairly indirect way of controlling the issue of central concern to users and investors: the rate of return on investment. Users want to be sure that investors are not earning an excessive return, while investors want to be sure they can earn an adequate return. If the returns prove too high or too low, a formula inhibits adaptation. At the same time, predicting rates of return likely to be earned under different tariff formulas can be very difficult; not only are the projects long-lived, but often substantial uncertainty exists about future traffic volumes or toll elasticities of demand.

Regulating rates of return directly creates its own problems, however. Determining the appropriate rate of return is difficult for new and untested industries. It remains to be seen how well the rate-of-return approach will work with privatized highways, even in the United States, which has long experience with such regulation in a variety of public utilities and other industries.

Prospects for Efficiency and Innovation

Balancing competitive, regulatory, and network issues is greatly simplified if privatization does, as its advocates contend, greatly reduce costs. Unfortunately, experience provides only limited support for the enhanced efficiency of privatization. In the cases reviewed here, only Cofiroute, the one surviving French private toll road company, displays direct, pertinent evidence. By achieving higher productivity of labor and equipment, Cofiroute apparently attained about a 23 percent cost savings on construction costs compared with the state enterprises constructing toll roads at the same time. Far less direct evidence on the relative efficiency of the two sectors is shown in the fairly detailed financial comparisons made of the costs of public versus private construction of the Dulles Toll Road Extension in Virginia; that study focused mainly on the financial costs of the two approaches, rather than possible direct differences in productivity, and indicated that there was very little to differentiate efficiency in the two sectors.[1]

Indirect evidence on the relative costs of the two sectors is, almost by definition, difficult to interpret and certainly ambiguous. For example, that private companies in Europe were more efficient in building limited-access highways than the public sector is certainly not proved by the survival rate for private toll road firms. France has the worst record, with three out of four of the original private French firms not surviving the energy market traumas of the 1970s. The private Spanish toll road builders did somewhat better. Only three of the thirteen private concessionaires established in Spain during the late 1960s and early 1970s had

1. José A. Gómez-Ibáñez and John R. Meyer, *Private Toll Roads in the United States: The Early Experience of Virginia and California* (Harvard University, Kennedy School of Government, Taubman Center for State and Local Government, December 1991).

to be taken over by government; another two, however, were absorbed, because of convenience or necessity, by other private companies. Of course, the French and Spanish private companies could still have been much more efficient than the public companies and had a lower survival rate only because they started later and lacked the deep pockets (access to taxpayer funds) of their public counterparts. A more accommodative Ministry of Finance in France in the mid-1970s might also have helped their survival, by allowing inflationary adjustment of private toll rates, as contractually specified in the original franchises. About all that can be said with any confidence is that any efficiency advantages the private sector might have had over the public were insufficient to survive the difficulties of the 1970s, especially in France.

There is also a good deal of evidence that both the French and Spanish privatization experiments were not structured to encourage the private sector to be efficient. The construction companies and banking concerns that were the principal backers and owners of the early private toll road companies may have been more interested in earning fees on the construction and financing arrangements than in creating and operating an efficient toll road. Since these fees were generally proportional to the monetary scale of the project, and initial equity requirements were often low, incentives to economize on construction costs could have been inconsequential, and sometimes even totally lacking. In both France and Spain, for instance, private toll road concerns undertook some extremely costly developments in relatively remote areas with low traffic densities, some places with densities so low and costs so high that it seems highly improbable that anyone contemplated any early achievement of operating profitability.

There are also conceptual reasons why private construction of toll roads might not be expected to achieve too many efficiency gains, particularly in the United States or other developed countries. Essentially, real productivity gains from privatization usually occur because the private sector can take better advantage of the learning curve or of scale or scope economies. In essence, a private vendor can often replicate activities at different locations as needs arise rather than reinvent the wheel as many local governments must do whenever they undertake these activities for themselves. The private vendors can therefore establish continuity, stability, and cumulative experience in a way that few government agency suppliers can. The private company may also be better able to afford highly specialized and costly staff services for finance, legal, design, and engineering needs. Many large construction companies often involved in highway construction, whether done publicly or privately, are truly international operations, sometimes with worldwide activities. Obviously, with operations at that scale, overhead can be spread thinly.

In many ways the main change effected by a private rather than a public toll road is in the nature of the competitive bidding or auctioning process that determines who is to do the construction and financing. In the public approach, separate bids are invited for the construction and financial underwriting (on reasonably rigidly specified contracts with the lowest bid normally winning). In the private approach the bidding is not unbundled or done separately; the franchise is a contract not just for the construction or underwriting but for both. With a private franchise, specification of what is to be done can be done by proposers, by the sponsoring government organizations, or on the basis of some interaction between the two sectors. In the United States, for example, private toll road proposals have mainly originated with the private vendors; in Spain government has largely specified what is to be done while in France a somewhat mixed approach has evolved.

Typically, a consortium seeking a private toll road franchise will include a large construction firm that could well be a bidder for the construction contract if the toll road were to be done by the public sector rather than the private. Similarly, the financial institutions that often become part of a consortium bidding for a private toll road franchise are likely to be candidates to underwrite debt financing for public road undertakings. In essence, then, with a private approach, the construction and financing are often done by the same people who would perform those functions if the road were built by a public authority. Since construction and financing costs, especially on a discounted present value basis, usually constitute a large part of the total cost of a high-performance highway facility, the opportunities for cost economies by shifting from a public to a private franchise are somewhat limited. This conclusion is reinforced because certain operating responsibilities, such as policing and environmental reviews, are likely to be at least monitored, if not actually performed, by the public sector even when a toll road is constructed and financed privately. In short, shifting from a public to private franchise on a toll road may not significantly alter the cast of players; and since many of the same people may be involved either way, costs may also not be too much altered.

This is not to say that there are no cost savings. Construction costs may be reduced, for example, simply because the private concessionaires' procedures for overseeing the construction may be more flexible than the public sector's procedures. This "flexibility," however, may be a disadvantage when the concessionaire is a contractor with little stake in long-term operations.

Moreover, innovative services and project designs may be a more important source of efficiency gains than reduced cost. Private provision of highways, like private provision of bus transit, does seem to make a difference for developing new techniques and services. Indeed, as with buses, a higher degree of innovation seems to be one of the strongest arguments for highway privatization. Many new and innovative proposals emerged in the U.S. experiences, both in California and Virginia. All of the successful California proposals had at least some innovative aspects. Ideas about how to vary toll rates with congestion levels and reduce toll collection costs as well as innovative alignments and construction techniques were prominent.

Environmental and Equity Issues

Environmental concerns are clearly of much greater import with highways than with bus transit, regardless of private or public sector ownership. Furthermore, these environmental concerns seem roughly proportional to the state of economic development in a society, or at least to the scale of automobile ownership. Simply put, the more automobiles per capita in a society, the more environmental concerns are likely to arise when highway construction is proposed.

Indeed, environmental concerns may be a greater obstacle to new highway building in the United States than financing or any other identifiable difficulties. Europe seems to be rapidly catching up with the United States, moreover, both in automobile ownership and environmental concerns. Not surprisingly, some of the most sensitive and hardest fought battles over highway construction have focused on construction of high-performance facilities in central urban areas. Environmental objections have stopped the construction of important high-performance highway links in several American cities, including San Francisco, New Orleans, and Boston. Similarly, the French are contemplating greater use of tunnels for new urban roads. Environmental concerns have thus become a serious obstacle to completing key linkages in highway networks, thereby becoming still another obstacle to attainment of the ubiquitous standardized network.

The U.S. experience with private toll roads suggests, however, that private companies sometimes may be more adept than their public sector counterparts in identifying the missing links and overcoming the associated environmental obstacles. Indeed, environmental opposition may create opportunities that the private sector is able to exploit. For example, the highest-ranked proposal in the California highway privatization competition was to build a highway over a river bed; to make the project economic, the vendor proposed a new and innovative construction technology to cut costs and building to specifications only sufficient to support automobile traffic, an exclusion that public highway planners might wish they could use but may feel politically constrained not to do. Similarly, the second-ranked project in northern California, held on standby in case the higher-ranked Midstate private toll road proposal does not happen, would mean completion of the San Francisco Embarcadero freeway, putting it underground to avoid the visual sight pollution that cut short earlier public efforts to complete this link; the private underground highway ostensibly would be financed from realizations on the real estate freed by removing unneeded elevated highway facilities.

Private owners may enjoy some advantages over public agencies in resolving siting problems, such as an ability to avoid the public spotlight until fairly late in the siting process, after many of the concerns of local residents and government regulations have been resolved. By contrast, a public agency usually must conduct a search for a new site openly from the start, so that local opposition has more chance to become mobilized and intransigent before concerns can be met. Private firms also may have more flexibility than public agencies in the compensation they can offer objectors and more incentive to compromise since their investors are so sensitive to the financial costs of delays. Many of these advantages were exploited by the backers of the Dulles Toll Road Extension, for example.

Any siting advantages of private firms may be offset, however, if the public is apprehensive that for-profit firms are less likely than public agencies to take their environmental and other community responsibilities seriously. Private involvement in highways may intensify siting problems by increasing local concerns that the new facility will bring too much or the wrong type of new development. Such suspicions are especially likely when, as is often true in the United States, the private toll road proposals are motivated in part by the development prospects they offer and are made financially possible by donations of right-of-way by the landowners who stand to gain.

In short, environmental concerns create opportunities as well as difficulties for prospective private toll road builders. The opportunities and the problems, though, are likely to accelerate as economic development proceeds. Indeed, in the less developed countries, the overriding goal may simply be to have the network constructed as quickly as possible.

Equity or distributional issues seemingly take a somewhat different form with highway privatizations than with bus transit privatizations. The usual losers in the privatization of bus transit are any riders who formerly rode at heavily subsidized fares and, to some extent, organized labor previously employed by public sector transit; the most common winners are taxpayers because of reduced payment of operational and capital subsidies. Highway privatizations are similar to transit in that taxpayers are usually counted among the winners because of reductions in indirect tax subsidies and because a tax-exempt public activity is converted into a taxable private activity.

To the extent that taxpayers gain, moreover, toll road users are sometimes losers. A private toll road will have to charge motorists more than a public toll road if it does not enjoy the same tax shields and is no more efficient. Toll road users might be better off with the private option, however, if there were compensating efficiency gains or if the public toll road would not be built as speedily or at all.

Highway privatizations, however, usually do not mean direct redistributions that are adverse to labor, simply because, as already observed, the same construction companies are likely to be involved whether the highway is built as a private or a public venture. Occasionally, though, and perhaps especially in the United States, labor may suffer some losses because of laws concerning allowable wage rates on public and private activities.

Highway privatization redistributions often have more to do with landowners than with any other group. Location of a high-performance highway's on or off ramps can often result in windfall gains being realized by certain property owners; generally those owners located at or near a ramp commonly experience large gains in the values of their properties while those at some distance from such sites do not. Obviously, the potentialities for such windfalls are much the same whether a facility is built by the private or public sector. A major difference, however, is that there may be more possibilities to strike deals when a private toll road is built in lieu of a public facility. For example, private highway providers may be less constrained and therefore better able to capture land rents or profits created by highway construction than public sector builders. The private sector may also have greater incentives and capabilities to build quickly so that greater capital-cost gains may be available, and bargainable with other parties.

Assessing the validity and universality of these possibilities and claims is difficult. It does seem plausible, though, that the bargaining combinations and possibilities are somewhat enhanced when a highway is developed by a private rather than a public operator. Furthermore, while transfers may be highly significant to the parties doing business, they do not seem likely to be serious issues for the body politic. Rather, the public is likely to be more interested in the tax consequences of private highway development and any effect that privatization might have on how speedily new facilities can be built and used. These equity issues do not seem to be much affected by the stage of economic development, although benefits to politically well-connected developers may be more important in less developed countries than in the developed.

Economic Development and Highway Privatization

In sum, highway privatization is a concept likely to prosper at only a few brief windows of opportunity. Privatization is likely to be most successful, politically and economically, at the take-off stage when an economy is just entering into modern consumerism. At that time, automobile ownership and usage will be surging and highway availability will be severely limited. In such circumstances identifying major highway linkages that are badly needed and readily financeable from toll collections is easy. Furthermore, automobile ownership will not be common enough to generate major environmental concerns. Any redistributions that might be effectuated by highway development are likely to be of little concern, easily subordinated to the overwhelming need for facility expansion.

Earlier, or relatively primitive, stages of economic development may have much the same configuration of policy concerns and political orientations but will lack one crucial element: the need for highways will not be as great, since automobile ownership rates are much lower, and thus self-financing projects will be much more difficult to find. Typically, a private investor undertaking a road development in a less developed country will have to wait longer for a traffic buildup to recoup investments. Self-financing from tolls will simply not be so plausible or easily attained.

At the other extreme, that of a highly developed consumer society, the obstacles are even greater. To start, identifying self-financing projects may be just as difficult as in very underdeveloped circumstances; the highway linkages that remain to be done are either developmental extensions of the existing system, into areas where traffic densities are likely to be somewhat low for some while, or highly costly augmentations of capacity in centrally located urban portions of the network where demand may be high but also concentrated into a few hours of the day. Furthermore, these extensions or augmentations will probably have to be built to relatively high and costly standards to be consistent with the rest of the national network. This combination of high cost and limited demands is unlikely to produce many viable and self-financing new toll road opportunities.

In developed economies, moreover, a host of new problems almost certainly will come to the fore, problems that usually do not arise, at least with the same intensity, at lesser stages of economic development. First and foremost among these additional problems are environmental concerns. The air pollution and congestion problems connected with high levels of automobile ownership are likely to be widely perceived and cause some anxiety. Redistributional issues may also be a bit more focused or understood and, since the highway system is already reasonably well-developed, not easily overridden by basic needs or requirements. As a consequence, the best argument for privatization in highly developed economies is likely to be that of breaking the bureaucratic mold and stimulating more innovation.

In short, highway privatization can be viewed as a delicate flower. It will bloom easily only under special circumstances, even though it might have considerable beneficial effect under many other circumstances as well.

Part Three

Profitability and Privatization: Airports and Urban Rail Transit

Chapter 12: Airport Privatization in Britain and the United States

THIS CHAPTER AND THE TWO THAT FOLLOW further explore the relationship between profitability and privatization by adding the experience of two other forms of transportation infrastructure: airports and urban rail transit.

As suggested by the experience of urban buses and high-performance highways, the prospects for privatization are greater for services or projects that require few subsidies, that is, where user charges or revenues will cover all or most costs. Airport and rail transit represent two extremes in profitability: airports being among the most profitable forms of transportation infrastructure and urban rail transit, one of the least profitable. An airport can be unprofitable if it is too large for the market it serves, but most major commercial airports usually recover all of their costs easily from charges to airlines, passengers, and commercial activities and concessions. Urban rail transit, with only a few exceptions, such as the Singapore and Hong Kong subways, covers only part of its operating expenses and none of its capital expenses out of the farebox.

Privatization is generally easier with more profitable types of infrastructure, but the relationship is not a simple one. Highly unprofitable facilities, such as rail transit in the United States, are usually very difficult to privatize because of the difficulty of convincing public officials or private enterprises of the merits of subsidizing the project. But highly profitable facilities are not necessarily easy to privatize either. Profitability may be regarded, fairly or not, as a sign that the facility operator might enjoy significant market or monopoly power, thus raising concerns about whether or how a private operator should be regulated. Furthermore, privatizing a highly profitable facility is often motivated by the public sector's desire to capture or recoup profits for other public purposes. Consequently, some facility users will face higher prices after privatization, and the prospect of price increases can mobilize opposition.

Airport Ownership and Operation

Small general or business aviation airports are often privately owned in many countries, but the airports offering regularly scheduled commercial air carrier services are usually public enterprises.[1] In a few cases private companies manage air carrier airports under contracts with the public authorities that own the airport. In the United States, for example, private contractors operate the airports in Burbank, California, and White Plains, New York. Beginning in the 1980s, several major public airports have awarded private companies concessions to build and operate terminals on their airports; terminal 3 at the Toronto Airport, opened in 1991, was the first important completed project of this type, and a similar project is under way at Istanbul's airport.

The only completely privately owned and operated major air carrier airports are in Britain. In 1987 the British government privatized the British Airports Authority, a public agency that operated London's three principal airports (Heathrow, Gatwick, and Stansted) and Scotland's four major airports (Prestwick, Glasgow, Edinburgh, and Aberdeen). Local governments, who own the rest of Britain's air carrier airports, were encouraged to privatize their airports and one has done so; in 1990 Liverpool sold a 76 percent interest in its airport to British Aerospace, an aircraft manufacturer who hopes eventually to expand the facility as a gateway and hub for Europe.[2] Britain has also permitted a private company to develop a small airport for short take-off and landing (STOL) aircraft at Docklands, an enormous office and residential development built on abandoned shipyards and wharves near the heart of London. Outside of Britain, Hong Kong and Athens are evaluating private concessions to build new airports,[3] and several U.S. cities have been considering selling or leasing existing air carrier airports.

Privatization of the British Airports Authority

British Airports Authority (BAA) was a state-owned corporation established in 1965 to operate four gateway international airports owned

1. This description of airport ownership patterns is based primarily on Robert W. Poole, Jr., *Airport Privatization: What the Record Shows* (Los Angeles: Reason Foundation, August 1990).

2. "Liverpool Airport Sold," *Public Works Financing*, July 1990, p. 6.

3. "Hong Kong Airport Privatization" and "RFP for Private Airport for Athens," *Public Works Financing*, July–August 1991, pp. 10, 11.

by the national government: Heathrow, Gatwick, and Stansted near London, and Prestwick in Scotland.[4] During the 1970s Scotland's three regional airports (Edinburgh, Aberdeen, and Glasgow) were transferred to the BAA. The BAA has a near monopoly on air passengers in Southeast England: in the mid-1980s its three London airports served 44 million passengers a year while the other airport in the area, Luton, had only 2 million passengers. The BAA's three London and four Scottish airports accounted for almost three-quarters of all air passengers and 80 percent of all international air passengers in the United Kingdom.[5] The remaining air carrier airports were mainly owned by local governments, for example, Manchester and Birmingham.

From its inception the BAA has been consistently profitable. The BAA was required by the government not only to cover its operating expenses but also to earn a reasonable return on any of the investments the government made. In 1987 the authority generated 48 percent of its income from "airside" or aviation services, which included fees charged to the airlines for landing aircraft, passenger terminals and baggage handling, and parking aircraft; the remaining 52 percent of BAA income was from the "groundside" or what the BAA called commercial services, which included rentals from duty-free shops, car rental firms, car parking, and other goods and services sold to passengers. These commercial services were usually operated by private firms rather than the BAA itself, with the concessions often awarded on the basis of competitive bidding. Although allocating the BAA's costs among activities was difficult, knowledgeable observers believed that the commercial services were cross subsidizing the aviation services.[6]

Airport capacity has long been controversial in London and Southeast England, where the demand for air services is substantial and the objec-

4. This account of the privatization of the BAA is based primarily on John Vickers and George Yarrow, *Privatization: An Economic Analysis* (MIT Press, 1988), pp. 354–66; David Starkie and David Thompson, "London's Airports: The Privatization Option," in John Kay, Colin Mayer, and David Thompson, eds., *Privatization and Regulation: The U.K. Experience* (Oxford University Press, 1986), pp. 210–20; Gellman Research Associates, *Analysis of Airport Cost Allocation and Pricing Options*, report to the Federal Aviation Administration (Washington, April 1990), pt. 2, chap. 5; and Poole, *Airport Privatization.*

5. The BAA's share of British air passengers would be slightly higher (Britain includes England, Scotland, and Wales while the United Kingdom includes Northern Ireland as well); U.K. Department of Transport, *Airports Policy*, report to Parliament, Command 9542 (London: Her Majesty's Stationery Office, 1985), p. 43.

6. Vickers and Yarrow, *Privatization*, pp. 357–58; and Starkie and Thompson, "London's Airports," pp. 210–13.

tions to aircraft noise, highway traffic, and land development are strong. From its inception, the BAA had supported the development of Stansted to relieve congestion at Heathrow and Gatwick. Community objections during the late 1960s forced the government to reconsider, but a three-year public inquiry (the Roskill commission) confirmed the choice of Stansted in 1971. In 1978 the Labor government essentially postponed the controversy over Stansted expansion by beginning a program to build a fourth terminal at Heathrow and a second terminal at Gatwick, which would enable the two airports to accommodate the expected growth in demand through 1990; no provision was made for the longer term. When the Conservative government took power in 1979 it continued the plans for the two new terminals but initiated a review of longer-term needs that led to another three-year public inquiry, concluded in 1984, over the capacity needed at Stansted.[7]

The Privatization Decision

A 1985 white paper on airport policy announced not only the government's proposal for privatizing the BAA but also its policy on future airport expansions near London. The white paper accepted the recommendation of the 1984 public inquiry that Stansted offered the only reasonable option for providing the capacity that would be needed by the mid-1990s. Improving Britain's regional airports to the north of London was not viewed as providing much relief for the Southeast because only about a quarter of the international travelers using London's airports were going to or from other regions of Britain, and even these airports were spread widely across the country. Expanding Gatwick or Luton would require controversial new runways while nearby defense bases could severely limit the airspace available for approaching aircraft. At Heathrow the capacity constraint was the terminals and not the airfield; adding a fifth terminal would not only aggravate the local community but would also require moving a sewage treatment plant.

Given these constraints, the 1985 white paper gave the BAA permission to develop Stansted to serve 15 million passengers a year in two phases of 7 million to 8 million passengers each. The government thought a fifth terminal at Heathrow might eventually be needed but would postpone its decision pending an evaluation of the feasibility and cost of

7. U.K. Department of Transport, *Airport's Policy*, pp. 3–4; and BAA, *1991 Annual Review* (London, 1991).

moving the sewage treatment plant. A limit of 275,000 aircraft operations a year at Heathrow, advocated by local community groups, was rejected, although the government announced that it might impose limits on airports in the future.[8]

The white paper listed several reasons for privatizing the BAA, including reducing the size of the public sector and promoting more innovative management.[9] Key functions would continue to be publicly provided, however. Britain's Civil Aviation Authority (CAA) would regulate safety and control the air space as before. The Department of Transportation would regulate security (antihijacking and terrorist measures), the secretary of state for the environment would continue to regulate aircraft noise and environmental impacts, while the public planning permission system would still govern major airport expansions.[10]

The most controversial aspect of the privatization was the government's decision to sell the BAA as a single company rather than breaking up the seven airports and selling them separately as some transport economists advocated.[11] Selling the airports separately, the government argued, would not significantly increase competition since most airlines strongly preferred locating at the largest airport in any area (in London, Heathrow) to provide the most convenient connections for their passengers. Moreover, the government had adopted rules to distribute traffic among London's airports. Charters were banned from Heathrow, and no new international carriers had been allowed to begin operations at Heathrow since 1977 (to preserve some space for domestic carriers at London's largest airport). Selling the London airports separately would reduce the government's ability to impose traffic distribution rules. Selling the London and Scottish airports separately was also rejected on the grounds that the Scottish airports benefited from common overhead functions performed by the BAA.[12]

The government would regulate the BAA, moreover, to prevent any monopoly abuses. Charges for aviation services (aircraft landing and parking and passenger handling) would be subject to the *RPI* minus *X* regulation that Britain had pioneered in its earlier privatizations of British Telecom and British Gas. Under the scheme, charges would be allowed to increase each year at the rate of the retail price index (*RPI*) less an

8. U.K. Department of Transport, *Airports Policy*, pp. 3–4, 15–24.
9. U.K. Department of Transport, *Airports Policy*, p. 44.
10. U.K. Department of Transport, *Airports Policy*, p. 47.
11. See, for example, Starkie and Thompson, "London's Airports."
12. U.K. Department of Transport, *Airports Policy*, pp. 44–45, 52.

estimate of the expected increase in productivity (*X*). For the BAA, the formula would be applied to the average charge per passenger for aviation services, but the company would be free to set rates for certain services (for example, per landing or per passenger) as it wished, as long as the average of all charges conformed to the formula. The formula would be revised every five years based on a review and recommendation by the Monopolies and Merger Commission (MMC), Britain's leading antitrust agency, to the CAA. For the initial five years, expected to begin in April 1987, the formula proposed was *RPI* minus one (*RPI* minus a 1-percent-per-year expected productivity improvement).[13]

The government argued that *RPI* minus *X* was superior to the rate-of-return regulation commonly practiced by public utility commissions in the United States (and applied to the Virginia toll road as described in chapter 10). Rate-of-return regulation requires detailed and often controversial investigations into the allowed investment base of the company, the reasonableness of the company's operating costs, and the appropriate rate of return on investment every time the company applies for a price increase. If the regulators over- or underestimated the needed rate of return, moreover, the company would have incentives to over- or underinvest. Although these same issues would have to be addressed in the five-year reviews of the *RPI* minus *X* formula, between these reviews the British regulatory scheme would be much simpler and less cumbersome.[14]

While the BAA charges for commercial services would not be governed by the *RPI* minus *X* formula, the MMC and the CAA would review the private company's commercial policies as part of the five-year reviews and could require changes. The CAA was also given broad authority to impose other conditions or regulations as needed. The CAA was empowered to insure that the privatized BAA honored the government's international obligations in aviation treaties, such as the requirement that all carriers be charged the same fees regardless of nationality and that fees be reasonably related to costs.

The Airports Act embodying the recommendation of the white paper was passed in 1986, and shares of the privatized BAA were sold to the public in July 1987.

13. In addition, the BAA would be allowed to pass on 75 percent of airport security costs imposed by the government. See Gellman Associates, Analysis of Airport Cost, pp. 5-10–5-21; and Vickers and Yarrow, *Privatization*, pp. 360–63.

14. For a comparison of the U.S. and British approaches see M. E. Beesley and S. C. Littlechild, "The Regulation of Privatized Monopolies in the United Kingdom," *Rand Journal of Economics*, vol. 20 (Autumn 1989), pp. 454–71.

Monopoly and Motives

The government's reasons for selling BAA intact could be challenged. Some observers questioned whether the government's traffic distribution policies were wise or necessary; it might be better for the market to determine which types of flights used which airports. (In fact, the government significantly relaxed its traffic distribution rules governing the three London airports in March 1991.)[15] Even assuming that control over traffic distribution was desirable, however, the government did not have to keep the BAA intact to do so; instead it might simply have dictated a minimum percentage of domestic flights at, say, Heathrow.[16]

Similarly, some competition would have occurred among the London airports even if Heathrow remained dominant. Charters, for example, are relatively footloose and less dependent on connecting traffic. Some scheduled traffic would probably have shifted to Gatwick or Stansted as well, if charges at Heathrow were too high. Heathrow also competes with Paris, Amsterdam, and other important European airports as an international hub and a gateway to Europe for tourists. Thus, while the government was probably correct that a breakup would not have eliminated the substantial market power of Heathrow, and therefore the need for some form of regulation to protect the public, a breakup would have greatly eased the burden on regulation.[17]

Similarly, the case for overhead cost economies by having one company was largely unproven. The four Scottish airports might have achieved sufficient scale taken together to achieve any such economies. Alternatively, any such scale economies might have been captured by contracting out certain staff functions. Finally, there was little evidence that the municipalities continuing to operate their own airports suffered substantial cost disadvantages.

Many observers speculate that the real reason for maintaining the BAA intact was to increase the price that the British government would receive for its privatization. The BAA was one of the first privatizations effected through the sale of stock to the general public, and the government was concerned that its offering would be fully subscribed. The three London airports would have greater market power, and thus greater value, if sold together rather than separately. The Scottish airports might have been difficult to privatize separately since they were thought to have been cross

15. BAA, *1991 Annual Review*, p. 8.
16. Gellman Associates, *Analysis of Airport Costs*, p. 5-22.
17. Starkie and Thompson, "London's Airports," pp. 214–15.

subsidized by the London airports in the past and in any event did not have an established record of profitability. In the absence of any other compelling reason for doing so, the most plausible arguments for keeping the BAA intact was to insure the success of the stock offering and increase the sale proceeds.[18]

The government's regulatory scheme also focused largely on potential monopoly in aviation services when a potential problem existed in commercial services as well. Breaking up the BAA would have increased competition for the aviation services but not necessarily for the commercial services, except to the degree that the two types of services were complements to one another. A 1985 review of BAA practices by the MMC noted that the BAA had not always encouraged the right kinds of competition in commercial services even when it was a state-owned corporation.[19] The publicly owned BAA usually awarded concessions on a competitive basis, but that did not translate into competition for consumers unless the BAA granted a number of competing concessions for each service. In fact, the MMC found effective competition in certain airport services, such as aircraft refueling and car rental, but not in others. A privatized BAA presumably would be even more interested in promoting competition *for*, rather than *between*, concessions. While the CAA had broad powers to control BAA commercial practices, commercial charges were not subject to the automatic *RPI* minus *X* controls.[20]

In the first four years after privatization, the profitability of the BAA increased steadily, with earnings rising from 21.0 pence per share in 1988 to 37.8 pence in 1991.[21] Most of the increase was on the commercial rather than the aviation side; commercial revenues grew from 52 percent to more than 60 percent of revenues in just three years.[22] The privatized BAA's investments have also been largely in terminals or on the commercial side; the BAA has opened several new hotels at its airports, for example, and in building new terminals and refurbishing old ones the BAA maximizes opportunities for shopping and other commercial activities. This response is understandable since commercial development provides the only opportunity for rapid revenue growth. Aviation revenue

18. Among those who make this argument are Starkie and Thompson, "London's Airports," pp. 214–15; Vickers and Yarrow, *Privatization*, p. 354; and Gellman Associates, *Analysis of Airport Costs*, pp. 5-22–5-23.

19. Vickers and Yarrow, *Privatization*, pp. 359–60.

20. Commercial activities would also be subject to Britain's ordinary antitrust laws, although Britain's laws are fairly weak.

21. BAA, *1991 Annual Review*, p. 1.

22. Poole, *Airport Privatization*, p. 13.

growth is limited to the increase in passenger traffic plus *RPI* minus *X*. The BAA argues that it has increased competition among commercial services at its airports, moreover, and has widely publicized its guarantee to consumers that prices at Heathrow and Gatwick shops and restaurants are fair.[23] Nevertheless, the growing dependence of the BAA on commercial activities suggests that it may be exploiting some market power in this area.

The *RPI* minus *X* regulatory scheme also has proven somewhat more difficult and controversial than British planners expected. The first five-year review, completed in 1991, was marked by a protracted and heated dispute between the BAA and the government over the appropriate rate of return on aviation investments. The BAA argued that its hurdle rate for its investments ought to be 15 percent, a rate it regarded as typical in the private sector. The MMC set an 8 percent rate of return as appropriate, however, on the basis that the BAA's effective monopoly near London made its investments far less risky than those of the typical private firm.[24] The CAA, which holds final authority in setting BAA rates, thought the MMC recommendation too generous and considered setting a 7 percent return. The BAA countered that such an inadequate return would force it not to invest in the proposed Heathrow-London high-speed rail link or the fifth terminal at Heathrow. In response, the CAA compromised on a 7.5 percent return; this figure required a lowering of the BAA's average aviation charges over the next five years, beginning in April 1992.[25]

Efficiency versus Transfers

Problems of monopoly and regulation would be more acceptable if the privatization of BAA held the promise of significant improvements in efficiency or investment. On these matters, the early record is somewhat murky. Some critics have argued that privatization should do little to increase economic efficiency at the airports since most activities there

23. BAA, *1991 Annual Review*, pp. 8–10.

24. Monopolies and Mergers Commission, *BAA plc: A Report on the Economic Regulation of the South-East Airports Companies (Heathrow Airport, Ltd, Gatwick Airport Ltd, and Stamstead Airport Ltd)*, report to the Civil Aviation Authority (Cheltenham, England: Civil Aviation Authority, 1991).

25. David Starkie, "Airport Investment: The Monopolies and Mergers Commission Report on BAA plc's South East Airports," *Journal of Transport Economics and Policy*, vol. 25 (September 1991), pp. 303–06; and "In a Holding Pattern," *Economist*, November 23, 1991, p. 20.

were always run by the airlines or commercial concessionaires and thus already effectively privatized. Furthermore, some of the efficiency gains claimed for the privatization of the BAA, such as the expansion of on-airport commercial activities, may largely be transfers. This judgment may be somewhat unfair, of course, since a privatized BAA might have more incentive than a public BAA to search out opportunities to add commercial activities at the airports. But any airport gain in such activities may be largely at the expense of off-airport merchants or hotels and thus more of a transfer of economic activity and rents from off-airport to on-airport sites rather than a net gain to society.

Other potential BAA efficiency gains appear not to have been exploited, largely because they might have been accompanied by politically controversial transfers. Investments in airfield or terminal improvements are still governed by the publicly controlled planning and environmental review process, for example, thus providing little scope for improvement from privatization. The privatized BAA, of course, might improve on designs or reject proposals that were uneconomic. Notably, however, the major terminal improvements made by the BAA during its first five years of private ownership were defined by past public inquiries.[26]

Similar political constraints seem to have limited the privatized BAA's use of pricing or other incentives to promote the more efficient use of existing airfield and terminal investments. The BAA's announced policy before privatization was that aviation charges would be set to cover long-run marginal costs. Long-run marginal costs are the incremental costs of landing an additional aircraft or processing an additional passenger over a period long enough that both investments and operating practices can be adjusted or optimized to the new level of output. Short-run marginal costs, by contrast, are the costs of accommodating an additional aircraft or passenger over a shorter period, when the investment in the facility is fixed. As explained in more detail in chapter 14, most economists agree that some form of marginal cost pricing encourages efficient use, with long-run marginal cost usually advocated when facility demand is growing and facility expansion is possible.

The BAA had made several noteworthy modifications to its charges during the 1970s to align them more closely with marginal costs. Investments were valued at replacement costs rather than historic costs, for example, to reflect the effects of inflation. The BAA shifted from aircraft landing fees that were proportional to aircraft weight, a practice still

26. Vickers and Yarrow, *Privatization*, p. 366.

common elsewhere, to fees based on the time it took to land an aircraft. From an economic perspective, this change made sense since runway time is scarce, and a light aircraft requires just about as much runway time as a heavy aircraft. Similarly, since excess runway capacity was available during the off peak but not in the peak, the BAA adopted surcharges for landing in the peak period, a practice not widely followed elsewhere.[27]

The BAA's aviation charges may have been set below long-run marginal cost before privatization, despite avowed policy. David Starkie and David Thompson calculated that long-run, cost-based fees for aircraft landing, aircraft parking, and passenger handling would have averaged 41 percent higher than actual fees at Heathrow in 1983 and 85 percent higher at Gatwick in the same year.[28] By adopting the RPI minus X cap on aviation fees for the privatized BAA, they argued that the government would perpetuate the preprivatization policy of underpricing aviation services.

Starkie and Thompson may have underestimated the prices needed to promote efficient airport use, and thus the appropriate price increases. At Heathrow and Gatwick, the possibilities for increasing capacity in the long run seem severely limited by opposition from neighboring communities or restrictions imposed by nearby military airfields. If expanding capacity is impossible, the long-run marginal cost of these facilities is essentially infinite; in such a context, the efficient price would be the market clearing price, presuming that this price at least covered the opportunity costs of using the airport for nonaviation purposes. Aircraft landing slots (a slot is the right to land at a specific time) and other scarce aviation services could be auctioned to the highest bidders. This practice would insure that those who valued those scarce slots the most would control them.

In short, achieving efficient airport use probably would have required large increases in aviation charges at Heathrow and Gatwick and probably a corresponding reduction in commercial revenue, exactly the opposite course from that adopted by the BAA and the government. From the government's perspective, dramatically raising aviation fees when the public was concerned that a privatized BAA might abuse its monopoly position could have been embarrassing. Large price increases would also have raised serious distributional issues and thus caused controversy. The

27. Gellman Associates, *Analysis of Airport Cost*, pp. 5-1–5-4.

28. At both airports fees for aircraft landing were less seriously underpriced than fees for aircraft parking or passenger handling. At Heathrow, for example, landing fees were twice as high as estimated long-run marginal costs while the other fees were well below costs; Starkie and Thompson, "London's Airports," p. 212.

current beneficiaries of low aviation service charges are most likely airline passengers or the airlines themselves, at least in less competitive airline markets. One study estimated that the airlines that have landing slots at Heathrow can charge their Heathrow passengers 10 percent to 13 percent more than passengers using alternative facilities.[29]

Privatization Proposals in the United States

In the United States the initiative for airport privatization has come from a few local governments rather than the federal government, in part because almost all major air carrier airports are owned and operated by local or regional governments. The national government is deeply involved in the operation of these locally owned airports, however. The Federal Aviation Administration (FAA) operates the U.S. air traffic control system, regulates some of the fees that airports can charge airlines and passengers, and provides capital grants to airports that in recent decades have covered about one-third of investments in airfield and terminal improvements.[30] Approximately one-half of all FAA activities are financed out of charges levied on airport users, the most significant by far being an 8 percent federal tax on commercial airline tickets.[31] Commercial airline operations are also thought to account for about half of FAA costs.

Albany and Other Proposals

The FAA has generally opposed privatization proposals, although local governments and market-oriented political interests have exerted intermittent pressures to change that policy. In the late 1980s the FAA's policy evolved in response to a proposal to privatize the Albany County Airport.[32] The Albany County Airport is the seventy-first busiest commer-

29. Study by David Thompson of the London Business School as cited by Douglas Cameron, "Airports Survey: Hidden Assets," *Airline Business*, December 1991, pp. 40–41. Of course, higher Heathrow charges might reflect higher costs or different traffic mixes (for example, leisure versus business) as well.

30. Congressional Budget Office, *Financing U.S. Airports in the* 1980s (Washington, 1984).

31. In addition, since 1990 commercial airline tickets have borne a 2 percent tax used for general government purposes.

32. Albany received about $2 million a year in FAA airport grants. This description of the Albany County proposal is based primarily on Esther Scott, "Privatizing the Albany County Airport," and "Privatizing the Albany County Airport: Epilogue," Teaching Cases C16-91-1024.0 and C16-91-1024.1, Harvard University, Kennedy School of Government, Case Program, 1991.

cial airport in the United States and provides the only air carrier service for an eleven-county area in the upper Hudson River valley in New York State, which includes the cities of Schenectady and Troy as well as Albany and has a population of about 1 million persons.

Albany County's primary motive for privatizing its airport was the financial gain from the airport's sale or lease. Cutbacks in state and federal aid to local governments and cost overruns on a new civic center had put the county in serious financial difficulty. Selling the airport was seen as a means of infusing cash, or at least preventing further financial drains from the airport, and thus helping to avoid a threatened property tax increase by the county.

Furthermore, Albany wanted to modernize its airport, and privatizing seemed a painless way to do so. Albany's terminal was too small and antiquated, without modern covered jetways. The federal rules governing public airports discouraged investment, however. Virtually every air carrier airport in the United States receives FAA airport aid.[33] All recipients are legally obligated to meet a number of FAA grant conditions, including a requirement that any fees or charges levied on airport users be used for the operation and improvement of the airport. This condition, strongly supported by the airlines, had been imposed to prevent recipients of federal airport aid from surreptitiously diverting such aid to nonairport purposes. The practical effect of the FAA requirements was to create an asymmetry in the risks and rewards of airport investment for local governments. On the one hand, if the investment was successful and all went as planned, airport revenues would service the bonds that the county issued to finance the improvements, but none of the profit could be taken "downtown" for other county purposes. On the other hand, if the improvements were not successful and airport revenues were not sufficient to meet the debt service, the county would be liable for the shortfall. Although the Albany County airport mainly operated in the black, it has run deficits at various times which the county has covered out of general revenues.

In early 1988 Albany County began exploring the idea of selling the airport for $4 million to the Capital District Transportation Authority (CDTA), a special purpose regional public agency that operated the area's public bus system but also had legal authority to operate other kinds of

33. The $2 million figure is Albany's annual allotment of FAA formula aid; in addition, Albany is also eligible for discretionary grants. For example, Albany received about $2 million a year in FAA airport grants. See Scott, "Privatizing the Albany County Airport," p. 2.

transportation services as well. By June 1989, however, a local private real estate firm, British American, generated a competing bid. British American controlled 400 acres adjacent to the airport and was concerned by tentative plans to build the new terminal on the opposite side of the airport from its property. British American teamed up with Lockheed Air Terminal, Inc., a subsidiary of the Lockheed Corporation that specializes in managing airports in the United States and abroad. The British American–Lockheed group proposed to buy the airport for $30 million, which provoked CTDA to raise its price to $24.25 million.

After some negotiation, the county effectively had two proposals. CTDA not only offered to buy the airport for $24.25 million but also promised to invest $117 million by replacing the existing terminal with an enlarged and modern facility and building a multilevel parking garage. British American–Lockheed offered to buy the airport for $30 million or to lease it for forty years for an initial payment of $25.5 million and annual lease payments rising from $500,000 in year 6 to $1 million in later years. The private group promised to spend $106 million on a new terminal and parking garage and to spend another $144 million to develop new hotel, office, warehouse, and retail buildings on and around the airport. The county's chief executive reportedly leaned toward the private proposal; the airlines that served Albany lobbied for the CTDA alternative, however, on the grounds that a private operator's profit motive would lead to higher fees.

By December 1989 the choice between the two proposals apparently became moot when the FAA, which had to approve the sale or lease of any airport, indicated that it was unlikely to approve either plan. Basically the FAA objected to both proposals because the purchase or lease payments would constitute an impermissible transfer of airport revenues to off-airport purposes. The FAA did not acknowledge in its decision that parking, restaurants, and stores in U.S. airports are often operated by private concessionaires who presumably take their profits off the airport. Apparently, the sale or lease of an entire airport, if not different in kind, was different enough in degree to concern the FAA.

Albany County abandoned its plan to sell or lease the airport in late 1990, after Congress passed a law authorizing airports for the first time to levy a passenger facility charge of up to $3 per passenger. Before that, airports could charge landing fees and lease gate and terminal space but could not collect a fee per passenger. With revenues from the new fee, Albany thought it could afford to build a more modest new terminal by itself. Moreover, the county now hoped to lease county-owned land across the road from the airport to a private parking lot operator for $15 million to $25 million for forty years. The terminal expansion would displace about half the existing parking spaces at the airport, thus creating a demand for the new parking facility. This parking scheme apparently was enough in keeping with practice elsewhere as to survive FAA review.

Several other local communities have also considered privatizing their airports, most notably Los Angeles. In Los Angeles the possibility of financial gain is as much or more of an incentive than any perceived need for modernization and expansion. The tax receipts from Los Angeles have been severely constrained by Proposition 13, the tax-limitation initiative approved by California voters in 1978. A 1992 report commissioned by the city council estimated that Los Angeles could realize $1.8 billion to $2.6 billion from the airport's sale. The estimate assumes that the concessionaire will be required to make almost $2 billion in terminal and other investments in the 1990s but will be allowed to raise airfield charges from their current low levels to about the average of other major U.S. airports.[34]

In April 1992, partly in response to the Albany failure, President George Bush signed a new executive order encouraging the privatization of state and local government infrastructure. It provides that state and local governments can free themselves of previous federal grant restrictions by repaying past federal grants. In calculating the repayment due, moreover, federal agencies are instructed to use the historic value of the grants and to deduct reasonable depreciation. This order seems likely to remove federal opposition to airport privatization, although its effects were still untested as of late 1993, but it may just shift the focus of activity by airlines and others to the local rather than the national level.[35]

34. In the study, airside charges were assumed to increase from an average of $2 per enplaned passenger to $6 per passenger; see Babcock and Brown and John F. Brown Company, *Los Angeles International Airport Privatization Study*, report to the City of Los Angeles (Department of Airports, May 1992). See also "LA Eyes Airport," *Public Works Financing*, December 1989, p. 2; and "LA Airport Study Discouraging," *Public Works Financing*, September 1990, p. 19.

35. "Special Report: Executive Order on Infrastructure Privatization," *Public Works Financing*, April 1992, pp. 5–10, "Airport Privatization Heats Up," *Public Works Financing*, March 1991, p. 2; and "New Look at Airport Privatization," *Public Works Financing*, December 1990, p. 11.

Monopoly and Regulation

As in Britain, a standard objection raised to airport privatization in the United States is the potential for monopoly abuse. Albany's airlines and some local business leaders raised this concern about the British American–Lockheed proposal, for example, and monopoly powers were also an issue for the FAA, although apparently not its primary concern.

The risk of monopoly is probably a less serious threat at most U.S. airports than at Heathrow or Gatwick. At a midsized airport like Albany, for example, some competition arises from other modes and other airports. Much of Albany's airline traffic is to New York City, for example, for which the car and train are reasonably competitive alternatives. In 1991 a flight from Albany to New York City (LaGuardia) took approximately one hour and cost between $149 and $238, while the train took 2.5 hours and cost $38; the train also provided direct downtown to downtown service likely to be more suitable for many business travelers.[36]

Many U.S. airports compete for hubbing flights, moreover, since the size and density of the United States make domestic hub-and-spoke networks a highly attractive airline strategy. Hub traffic is notoriously mobile and sensitive to the services offered and fees charged by airports; for example, several cities in the southeast could easily replace Charlotte as USAir's regional hub; and Denver, Chicago, Dallas–Fort Worth, Minneapolis, St. Louis, and other cities all compete as transcontinental hubs. Some studies have shown that when a single airline dominates an airport in the United States (as Northwest does at Minneapolis, for example, or USAir at Pittsburgh), it can charge higher fares for tickets to and from that airport.[37] The airline can charge higher prices only to passengers whose ultimate origins or destinations are the "dominated" hub, and thus presumably the airline's captive, but not to passengers who are making connections at the hub and presumably have more choices. An airport can not easily make the same distinction in its landing fees, however, since an aircraft typically will carry local and connecting passengers.

36. Scott, "Privatizing the Albany County Airport," p. 4, n. 5.

37. See, for example, Elizabeth Bailey and Jeffrey R. Williams, "Sources of Economic Rent in the Deregulated Airline Industry," *Journal of Law and Economics*, vol. 31 (April 1988), pp. 173–202, esp. pp. 183–187; or Steven Borenstein, "Hubs and High Fares: Airport Dominance and Market Power in the U.S. Airline Industry," Discussion Paper 278 (University of Michigan, Institute of Policy Studies, March 1988).

The FAA also has broad legal authority to insure that an airport that has received federal aid will "make (their) airport available for public use on fair and reasonable terms and without unjust discrimination to all types, kinds, and classes of aeronautical uses."[38] Furthermore, local governments could retain rights to regulate airport charges as one of the terms of their lease or sale.

Transfers and Efficiency

The most fundamental obstacle to airport privatization in the United States is probably the concern of airlines, and other aviation groups, that they would face higher airport charges, even where privatized airports did not enjoy significant monopoly power. Indeed, the system of FAA regulations and grant conditions is designed in many ways to keep airport charges low, as an aid to developing the industry, and perhaps reflecting the political influence that some aviation interests have historically enjoyed.

As one consequence, if the BAA's airport charges were below long-run marginal costs before privatization, U.S. airports' charges are probably significantly more so.[39] Even before privatization, the BAA was required to earn a reasonable return on its investments, while the FAA essentially prohibits any airport from earning a return on the federal airport grants it receives by prohibiting the use of airport revenues for nonairport purposes. Furthermore, while the BAA has valued its investments at replacement costs since the 1970s, U.S. airports normally value their assets at historic costs.

Perhaps the most important factor leading to underpricing in the United States is the accounting used for setting U.S. airside charges.[40] Two accounting approaches are widely employed. Approximately 42 percent of U.S. airports use what is known as compensatory cost for setting charges to airlines. Under this scheme airlines are charged only the direct costs of the facilities and services they use. Not only are investments valued at historic costs, but facilities that arguably might be the

38. As quoted in Scott, "Privatizing Albany County Airport," p. 10, n. 10.

39. For an eloquent description of the factors leading to low U.S. airport charges see Frank Beradino, "Airport Privatization," presentation to the AAAE/ABA Seminar on Airport Law, Minneapolis, October 3, 1990; for a comparison of U.S. and British airport practices see Gellman Associates, *Airport Cost Allocation*, chap. 6.

40. The two approaches are described in Congressional Budget Office, *Financing U.S. Airports*, pp. 19–33.

airlines' responsibility, such as public areas in terminals, are often excluded from the airlines' costs. The remaining 58 percent of U.S. airports use residual cost; under this scheme the airport deducts its commercial revenue (for example, lease payments from shopping, car parking, and other concessions) from its total costs and charges the airlines only the residual.

These accounting conventions encourage the cross subsidy of the airport's aviation services with revenue from nonaviation or commercial activities. As one result, commercial revenues account for a much larger share of total airport revenues in the typical U.S airport than they did for the BAA before privatization.[41] Moreover, airport landing fees and other charges to airlines appear slightly lower in the United States than in Britain as illustrated by the comparison in table 12-1 of charges for international flights at Heathrow and U.S. airports in 1988. Heathrow's charges were 11 percent higher than the average charges in the twenty U.S. airports surveyed. Of the six major U.S. airports with numerous cross-Atlantic flights shown in table 12-1, all but one, Kennedy Airport in New York, charged fees much lower than Heathrow. Moreover, Heathrow's charges are probably far below both its long-run marginal costs and the market clearing price for its scarce capacity.

U.S. airport charges are not structured to encourage efficient use of scarce runway or terminal capacity. While the BAA switched to landing charges based on runway time in the 1970s, landing charges based on aircraft weight are still the norm at U.S. airports, thus seriously undercharging lighter aircraft. And unlike Britain, peak-hour differentials are also extremely rare in U.S. airports and usually apply only to general aviation rather than commercial air carriers.

The FAA may have increased airline opposition to any change in these airport pricing practices by its decision in the late 1980s to grant the incumbent airlines the rights to sell their landing slots at the nation's four most congested airports (Kennedy and Laguardia airports in New York, O'Hare Airport in Chicago, and National Airport in Washington, D.C.).[42] Landing slots at these U.S. airports had been allocated by sched-

41. How much larger is not clear; a recent study by Gellman Associates reports that commercial activities account for 75 percent of the revenue at a typical U.S. airport versus 50 percent in the BAA. The Gellman study reports no sources, however, and an older Congressional Budget Office study suggests that the commercial share of U.S. airport revenues was only perhaps 60 percent in the 1970s. See Gellman Associates, *Airport Cost Allocation*, p. 6-4; and Congressional Budget Office, *Financing U.S. Airports*, p. 35.

42. Vlad Jenkins, "The Department of Transportation and Airport Landing Slots," Teaching Case C16-87-781.0 (Harvard University, Kennedy School of Government, 1987).

TABLE 12-1. Representative Landing, Parking, and Passenger Charges, Including Terminal Navigation and All Passenger Charges and Taxes
1988 dollars unless otherwise noted

Airport	*DC-9-30*	*A300B2*	*747-200B*	*Weighted index*[a]
London(LHR)[b]	1,022	2,339	3,480	111
New York(JFK)[c]	1,134	2,763	4,451	130
New York(LGA)[c]	599	1,523	3,256	77
New York(EWR)[c]	543	1,443	2,820	70
Chicago[c]	786	2,200	3,433	98
Los Angeles[c]	257	722	1,155	32
Miami[c]	599	1,594	2,559	73

SOURCE: "Airports Restrained, Airlines Unimpressed," *Avmark Aviation Economist*, vol. 5 (December 1988), p. 5.
a. The index is based on international flights, operating at 65 percent passenger load factors. Based on twenty airports, 100 is the average of all airport charges.
b. Weighted average of peak and off-peak charges.
c. Includes $3 ticket tax paid by departing passengers.

uling committees of the airlines; the committees were expected to reach unanimous consensus, with the FAA intervening only in deadlocks. As air traffic grew in the early 1980s the scheduling committees at these four airports could not reach agreement on applications for slots by new airlines. Rather than rationing demand by allowing landing fees to rise to market clearing levels, however, the FAA instead allowed the airlines to buy and sell their slots. The existing holders of the slots received a windfall in their sudden receipt of marketable assets (landing slots). While the FAA denies that the airlines have legal property rights to the slots, any increase in landing charges will obviously reduce the market value of these slots and thus is likely to be opposed by the airlines.

A shift to airport charges based more on costs or market values thus would clearly be a more radical reform in the United States in the 1990s than in Britain. The efficiency improvements from the shift might also be greater in the United States, but so would the opposition from the well-entrenched interests that benefit from the current system.

Lessons

In sum, the experience with airports suggests that profitability, while helpful, does not necessarily guarantee the prospects for privatization. In the first place, the possibility of large profits may raise the specter of monopoly abuse and related problems of regulation. This fear of monopoly has probably been exaggerated, especially as it pertains to U.S. airports, since they often appear to face formidable competition from other airports and other modes of travel. Nevertheless, the fear persists, and in some places perceptions may be as important as reality.

The prospects of large profits may also affect the motives of privatization proponents and the intensity of their opposition. In the United States an important motivation for airport privatization proposals has been the desire of local governments to capture or cash in some of the potential profits and use them for nonairport and more immediate purposes. Even in Britain, the prospect of increasing the proceeds from the sale of the BAA is alleged to have led the government to sell its seven airports as a group rather than separately, thus exacerbating the monopoly potential and reducing the possible efficiency gains from private ownership. The higher profits expected from privatized airports are based partly on the expectation of higher charges for airport use; this serves to mobilize those who benefit from the present system to oppose change. The main direct effect of airport privatization in the United States, in regard to the distribution of wealth, would probably be a transfer from travelers and airline stockholders to local government taxpayers.[43]

The other major redistribution likely to result from airport privatization is a heightening of the proportion of total airport revenues realized from commercial rather than airside services. Almost every principal public airport seems to have unexploited possibilities for the expansion of retail and other service activities, as confirmed by the early British experience, suggesting, in turn, a strong possibility, or at least temptation, to exacerbate the tendency to cross subsidize airside activities from commercial (groundside) operations. Of course, to the extent that higher airside fees are also charged because of privatization, then the net effect may not be a higher cross subsidy but rather higher net income realizations from airport operations, possibly made available for other government activities through higher privatization sale prices, higher taxes, or higher lease fees. Almost certainly, too, more funds should be available for airport improvements and investments.

In short, airport privatization opens up a plethora of possibilities and political complexities: important new sources of local government revenues, higher property tax revenues, substantial redistributions, expanded investment programs, and so on. The possibilities have been so rich and intriguing, and so hemmed in by controversy and legal technicalities, as to escape easy resolution, except in Britain and a few other isolated cases.

43. The competitive nature of U.S. airlines virtually insures that any landing fee increase or decrease will almost immediately be reflected in ticket prices.

CHAPTER 13. The U.S. Experience with Private Rail Transit

with ARNOLD M. HOWITT and ALAN D. WALLIS

IF AIRPORTS ILLUSTRATE THE PROBLEMS of privatizing profitable undertakings, urban rail transit illustrates the even larger difficulties with privatizing unprofitable activities.[1] Urban rail transit includes subway, streetcar, cable car, monorail, and other fixed guideway systems designed with relatively short distances between stations to serve both commuting trips to the center of a metropolitan area and shorter trips within the metropolitan center. Urban rail is distinguished from its close cousin, the commuter railroad, in that the latter has much longer distances between stations and is designed primarily to serve longer commuter trips from the outer suburbs to the metropolitan core. Commuter railroad services often share tracks used for intercity passenger or freight service while urban rail transit almost always enjoys the exclusive use of its right-of-way.

Economic and Ownership Trends

History offers little support for the proposition that urban rail transit systems can be run profitably, whether publicly or privately owned.[2] The basic economics of rail operations are fundamentally much less adaptable than those for buses, mainly because of the substantial investment in the fixed guideway and larger vehicle (that is, train) sizes. Bus service can often share the streets with automobiles, and a dedicated bus lane, if necessary, is typically far less costly to build than a dedicated rail line. An efficient bus operation can be established with vehicles carrying 20 pas-

1. The case studies in this chapter are drawn from José A. Gómez-Ibáñez and others, *Prospects for Privatizing Urban Rail Transit: Lessons from Three Case Studies*, report prepared for the Urban Mass Transit Administration (now the Federal Transit Administration) (Harvard University, Kennedy School of Government, Taubman Center for State and Local Government, December 1991).

2. One astute observer once said that "the private sector does not invest in rail transit facilities, in particular, because this form of transportation has been failing a market test with respect to new investment since the Panic of 1907." George W. Hilton, *Federal Transit Subsidies* (Washington: American Enterprise Institute for Public Policy Research, 1974), p. 117.

sengers or fewer, while an efficient rail transit operation often requires trains with a passenger capacity of 500 to 1,000 passengers. As a consequence, travel corridors in smaller metropolitan areas and from the outer suburbs of larger metropolitan areas rarely have sufficient volumes to support profitable rail service.[3] Workplaces and residences are suburbanizing or decentralizing in virtually every metropolitan area in the world, moreover, further reducing the market for urban rail transit.

Many rail transit and commuter railroad systems were originally built by the private sector in the last half of the nineteenth and the first decades of the twentieth centuries. Virtually all of these were subsequently taken over by government as competition from the auto and bus and suburbanization reduced rail patronage and profitability. Some commuter railroads still remain in private hands, but they are usually operated under contract to local government and heavily subsidized, as in the United States, for example. The principal exceptions are in Japan where approximately a dozen major commuter railroads serving its largest metropolitan areas are privately owned and profitable.

Several countries have begun to privatize their public railways, although usually intercity passenger and freight rather than commuter or rail transit lines are involved. Japan broke up Japan National Railway into six separate private companies in 1987 with the intention of weaning these new companies from subsidies and selling them off in five years. The United States set up a publicly owned but "for profit" corporation, Amtrak, to run its intercity passenger service in 1970, but it was still dependent on public subsidies twenty years later. Argentina began to sell the commuter railways serving Buenos Aires to the private sector in 1992. However, these privatizations of existing public rail lines often depend on the fact that the historic (but still valuable) investments in the rail lines have been largely written off by the public sector with no need to be recovered by the new private owners.

Private investment in *new* urban rail lines is therefore extremely rare, although there are some instances. Several of Japan's profitable private

3. See John R. Meyer, John F. Kain, and Martin Wohl, *The Urban Transportation Problem* (Harvard University Press, 1965), pp. 299–306; Theodore E. Keeler and Kenneth A. Small, *Automobile Costs and Final Intermodal Cost Comparisons*, vol. 3, *The Full Costs of Urban Transport*, ed. Theodore E. Keeler, Leonard A. Merewitz, and P. M. J. Fisher (University of California at Berkeley, Institute of Urban and Regional Development, 1975); and J. Hayden Boyd, Norman J. Asher, and Elliot S. Wetzler, *Evaluation of Rail Rapid Transit and Express Bus Service in the Urban Commuter Market*, report prepared by the Institute for Defense Analyses for the U.S. Department of Transportation (Washington, 1973).

commuter railroads built new branch lines or extensions during the 1970s and 1980s ; Walt Disney built monorails in its amusement parks, and a real estate developer built a short rail line for internal circulation in a major new office and retail center at Las Colinas near Dallas, Texas. In large measure, though, new rail transit lines have been the province of the public sector for many decades.

Nevertheless, proposals for new private urban rail transit systems surfaced in several countries during the late 1980s and early 1990s. In 1991 and 1992 Thai officials signed concession agreements for three different new, private elevated rail transit lines in Bangkok, although none had begun construction as of late 1993.[4] Backers of two of the three lines argue that they will be profitable from fares alone given Bangkok's extraordinary traffic congestion. The third project combines an elevated rail line with an elevated expressway and retail developments around the stations; backers hope to use profits from the real estate ventures to offset projected losses on the rail line. Several serious proposals for private rail lines also emerged in the United States in the mid-1980s, and these plans are the focus of this chapter.

The explicit policy of the U.S. federal government from the early 1960s to the early 1980s was to provide financial support to help local governments build, acquire, and operate mass transit systems of all kinds, buses and rail. Policies of the U.S. Federal Transit Administration (FTA) made it much easier for a local government to qualify for federal transit grants and aid if the local transit system was publicly rather than privately owned.[5] In practice, this meant converting most of the remaining private urban transit companies to public ownership.

A significant portion of the FTA's capital grants also went toward development of rail transit, reviving long-dormant interest in such systems. Since up to 80 percent of rail capital costs could be covered by federal grants, local governments and agencies were often tempted to pursue inefficient investments. Rail systems were built in areas lacking

4. For background on these three projects see "Skytrain Touches Down after a Long, Rough Ride," "Hopewell Hits on High-Rise Train/Retail Network," and "Office Builder Turns to Light Rail Transit for Relief," in *Public Works Financing International*, July 1992, pp. 7–9, 9–11, and 11–12.

5. The Federal Transit Administration was called the Urban Mass Transportation Administration until 1991; for simplicity, its current name is used throughout. The 1964 act did not explicitly preclude private sector participation in federally sponsored programs, but federal grants could only be provided to public agencies. As a result, a public agency would have to receive the grant and own any equipment purchased with the grant, but it would then lease this equipment to a private company for a nominal cost.

adequate density to generate ridership sufficient to cover operating costs, let alone capital costs.

These inefficiencies did not escape the FTA or its critics. In 1975 the agency began requiring that all grant applicants demonstrate that they were effectively utilizing existing highways and transit facilities as a prerequisite for requesting funds for new rail facilities. In 1978 the FTA introduced policies approving new rail systems only in stages, with preference given to segments serving the densest urban corridors.[6] Nevertheless, massive capital grants for the development of new systems continued and by 1988, $12 billion had been spent by the FTA to support investment in new rail facilities (out of $37 billion total appropriation for capital assistance to the transit industry).

Despite these reforms, the performance of FTA-funded rail projects continued to be disappointing. With higher costs and lower ridership than forecast, these new systems required much higher operating subsidies than expected.[7] Partly in reaction to these adversities, the FTA, beginning in the early 1980s, significantly changed course and started encouraging greater private sector involvement in transit. New budget realities dictated the FTA's shift: federal subsidies for urban transit operating expenses were cut by 40 percent (in constant 1981 dollars) between 1980 and 1988.

The message was not lost on local governments. With federal transit contributions declining and growing resistance to new local taxes, local governments were already seeking private participation in transit service and infrastructure. In New York City, for example, Citicorp, to gain approval of a new office building in Long Island City, agreed to pay for construction of a new subway station. An increasing number of communities required developers of new commercial office space to help mitigate the traffic congestion that these developments would cause. In some cases, these measures included private forms of transit, such as shuttle buses or van pool programs that were operated and subsidized by the developer or tenants. In states that permitted it, special fees, usually called exactions, also helped generate capital for transit improvements.[8] Several local public transportation authorities contracted with private firms for

6. John R. Meyer and José A. Gómez-Ibáñez, *Autos, Transit and Cities* (Harvard University Press, 1981), p. 50.

7. Don H. Pickrell, *Urban Transit Projects: Forecasts vs. Actual Ridership and Costs* (Cambridge, Mass.: Transportation Systems Center, October 1989).

8. Alan A. Altshuler and José A. Gómez-Ibáñez, *Regulation for Revenue: The Political Economy of Land Use Exactions* (Brookings, 1993).

services on routes that had formerly been served by publicly owned companies, and they often discovered that the private operators could provide a comparable or better-quality service at lower costs (as described in chapter 5).

There were also some encouraging signs of private sector interest in rail transit. The new Las Colinas rail line and the Disney monorails were private transit systems serving wholly private developments. Nevertheless developers found these fixed guideway systems compatible with an overall objective of achieving profitability, suggesting that the mode was not inherently inappropriate for privately provided public transit.

While several proposals for new private rail transit lines that would serve the general public were broached in the United States during the 1980s, only three appeared to have some promise of success (although none had begun construction as of the early 1990s). In all three cases the private investors proposed significant funding for a new fixed guideway service, not just limited and specialized contributions (such as building one station or paying transit impact fees or exactions). Services would be available to the general public and were not to be built entirely within a single land development, owned by a single company (as in the Disney and Las Colinas cases). All had been under development for several years, and their promoters had spent a great deal of money in feasibility and planning studies. These three—located in Boston, suburban Washington, D.C., and Orlando, Florida—together offer a variety of experiences and insights about the problems of rail transit privatizations.

Boston's World Trade Center Monorail

In 1985 developers of the Boston World Trade Center opened a convention and office complex in the wharf area of south Boston, nearby Boston's downtown financial district. They then proposed building another 2.8 million square feet of offices adjacent to their existing buildings. As part of their expansion, they offered to build a $26 million monorail to connect the new office buildings with an existing rail transit and commuter rail station (South Station) less than a mile away.[9]

Originally, the developers intended to build and operate the monorail at their expense, with no fares charged to the passengers and no direct subsidies from public authorities. The proposed line would help encour-

9. For a more detailed description of this porposal and its subsequent experiences, see Gómez-Ibáñez and others, *Prospects for Privatizing Urban Rail Transit*, pp. 13–63.

age development in a waterfront area that the city and state had slated for renewal. The Boston World Trade Center developers saw their proposal as a win-win proposition: they would get a marketing tool for their 2.8 million square foot expansion project, and the public would receive an attractive, high-technology transit system that would help solve the transportation problems associated with development of the wharf area known as Fort Point.

By late 1986, encouraged by state and local officials to refine their ideas, but getting no tangible backing, the developers sought help from the federal government. The FTA saw the Boston World Trade Center monorail as a potential national model for private investment in transportation infrastructure, and thus the Boston project became a policy priority for the agency. To insure that such a showcase project could go ahead, the FTA offered to pay for detailed engineering plans and, ultimately, to finance up to one-half of the project's capital costs.

Despite FTA backing, the World Trade Center proposal ran into serious difficulties during 1987 and 1988. To maintain control over the decisionmaking, the Massachusetts secretary of transportation insisted that the state undertake its own systematic study of Fort Point transit options before proceeding. As that study progressed, it became increasingly clear that top Massachusetts officials had serious doubts about the private monorail plan. Moreover, another major developer in the area opposed the project, mainly because of an apparent belief that it would bestow an unfair competitive advantage on the World Trade Center site. Nor did any other landowner step forward to endorse it.

State officials argued that the monorail could serve only a fraction of the ultimate transit demand in the area, a view fueled by the ebullient real estate markets of the 1980s. In the state's view, the monorail linked up very imperfectly with the rest of Boston's transit system. State officials also did not believe that an elevated structure was politically acceptable at a time when the state was going to great expense to eliminate other elevated transportation facilities in Boston (in particular the downtown Central Artery Highway). Finally, they were skeptical about the developers' estimates of ultimate cost, the speed with which public approvals could be obtained and construction commenced, and assurances that potential operational problems could be overcome.

The conduct of the state study angered both the developers and the project's supporters in the FTA. They saw the state as playing political games rather than fairly assessing the developers' plan. They believed that state policymakers were philosophically hostile to serious private involvement in the provision of transit and foolishly ready to turn away private capital investment in the vain hope that the federal government would pay for a far more expensive transit alternative. For their part, state officials saw the developers pursuing their own short-term financial interests, while they viewed the FTA as more concerned with a national ideological agenda than with the long-term needs of the Boston transportation network.

As the decisionmaking process dragged on into 1988, the economic climate turned chilly, especially for commercial real estate. This finally killed all immediate plans for major development in the area. The World Trade Center developers abandoned the monorail proposal and deferred plans to proceed with World Trade Center expansion. Indeed, the development partner that had taken the lead in pressing for the monorail sold its interest in the entire World Trade Center. By late 1988 the Boston monorail offer was therefore moribund.

The Dulles Airport Connector

In 1983 a local private developer suggested that a coalition of local developers pay for construction of a sixteen-mile rail line to connect Dulles International Airport in the Virginia suburbs of Washington, D.C., with the nearest subway station on the D.C.'s Metrorail system. In 1984 another set of backers proposed that the line be privately built and operated by their company, Dulles Area Rapid Transit (DartRAIL).[10]

While completing the rail link would help make distant Dulles become the major airport it was originally intended to be, DartRAIL also promised to help meet the transportation needs of Fairfax County, Virginia, one of the fastest growing areas in the Washington region. Fairfax County is bordered on the west by Dulles Airport and on the east by I-495, better known as the Washington beltway. Tysons Corner, at the eastern edge of Fairfax County, and on the beltway, contains almost 40 million square feet of office and commercial space, the largest concentration in the metropolitan area outside of downtown Washington. Further west, and about half-way between Dulles and the beltway, is Reston, a major new town started in the 1960s with 10 million square feet of office and commercial space. Just beyond Reston are major concentrations of office and commercial space in the city of Herndon.

10. For a detailed account of this case see Gómez-Ibáñez and others, *Prospects for Privatizing Urban Transit*, pp. 65–109 .

From the outset DartRAIL's promoters realized that they had to get financial support from other private entities or the public sector to make their project viable. DartRAIL's plans changed several times as its different requests for outside support proved impractical or politically unacceptable.

DartRAIL's First and Second Plans

DartRAIL's first proposal, made in February 1985, was to build a light rail line in the median of the Dulles Airport Access Road (which ran nonstop from the Washington beltway to the airport). Passenger fares would cover operating costs but not the estimated capital cost of $300 million. To raise the capital, DartRAIL proposed that the Federal Aviation Administration (FAA), then the owner of Dulles Airport, lease it 600 acres of land at the airport for little or no cost. DartRAIL would, in turn, sublease the land to developers for office buildings, hotels, and other projects, and use the profits from the subleases to build the rail line. The FAA, which had not been consulted in advance, was lukewarm about the requested donation and concerned that the proposed real estate development on airport land might interfere with future airport improvements. Local government officials also worried that DartRAIL's finances were not credible, and that they might be left with a significant liability if DartRAIL failed to complete the project or could not operate it.

Reacting to these concerns, DartRAIL advanced a new plan in March 1989.[11] The new plan called for a heavy rail line in the median of the Dulles Airport Access Road that would be fully compatible with Metrorail and provide transfer-free access between downtown Washington and Dulles Airport. The new line would also have six intermediate stations to serve Fairfax residents commuting to downtown or to jobs in Fairfax (these commuters were expected to constitute more than 80 percent of DartRAIL riders). Tysons Corner would be served by a four-mile elevated monorail loop, with as many as nine stations, that would connect with the Tysons Corner station of the heavy rail line. Road improvements, parking lots, and feeder bus drop-off points would also be built to provide ready access to the system by car, van, or bus. The estimated costs were considerably higher than those of DartRAIL's 1985 light rail proposal. Construction costs (in 1990–93 dollars) were esti-

11. Dulles Access Rapid Transit, Inc., *An Action Program for Comprehensive Transportation Improvements in the Tysons Corner/Dulles Corridor Area* (Vienna, Va.: March 1989).

mated to be $532 million, including $283 million for the heavy rail line (and twenty-eight cars), $77 million for the Tysons Corner monorail, and $172 million for construction financing and reserves.

The plan assumed that Fairfax County, the airport, and Washington Metropolitan Area Transit Authority (WMATA) would assume some important responsibilities. The county would operate and pay for the feeder bus service to the rail stations. The airport would be responsible not only for providing the right-of-way in the access road median and at the airport, as had been true in the earlier plan, but also for building the Dulles Airport station (whose costs were not included in the $532 million figure). WMATA, the Metrorail operator, would agree to allow DartRAIL to operate on its tracks in return for a share of the revenues and to maintain DartRAIL's cars at its yards for a reasonable fee.

In this second plan, DartRAIL changed its strategy for financing capital costs, which it again predicted could not be recovered from passenger fares. Instead of asking the airport to lease its land for real estate development, DartRAIL would ask private real estate developers along the line to pay. Virginia had recently passed a law allowing the creation of Transportation Service Districts (TSDs) to finance special transportation infrastructure improvements. If owners representing a majority of the value of commercial and retail property agreed, their area could be declared a TSD and a special tax of up to $0.20 cents per $100 of assessed valuation could be collected to finance needed improvements. DartRAIL proposed the creation of a TSD at Tysons Corner and along the Dulles–West Falls Church corridor. The Northern Virginia Transportation Commission (NVTC), a state commission with the authority to issue tax-exempt bonds, would issue the bonds secured by the TSD revenues and would own the system.[12] DartRAIL would operate the system under contract to the NVTC.

Fairfax County officials raised objections to DartRAIL's new plan as soon as it was released. In response to the earlier DartRAIL proposal, Fairfax County had secured an FTA grant to do its own study of mass transit options, public and private, in the Dulles corridor. A draft of that study, circulating while DartRAIL's second plan was formally presented, argued that a bus alternative would be far more cost effective than a rail

12. Under the Virginia law, a TSD does not have the authority to issue its own bonds; the state would issue the bonds on its behalf, secured by a contract that gave the state the TSD's future revenues. Additional security was provided by allowing bond holders to place a claim against Virginia's highway trust fund in the event TSD revenues were less than expected.

line.[13] Indeed, under regulations governing FTA-funded studies, the county's consultants had to include a low-cost, bus-based transportation systems management (TSM) option as an alternative to the rail proposals. The consultants compared two bus options with six rail alternatives, including some closely resembling DartRAIL's proposals.

The consultants argued that none of the alternatives would make much of a dent in local traffic congestion, and that the rail options would attract only a few thousand more riders than the TSM option. Moreover, TSM was significantly cheaper than any of the rail alternatives. TSM would have capital costs of $72 million (in 1988 dollars), for example, while the cheapest rail option (light rail) had a capital cost of $562 million. All the alternatives would recover only 50 percent to 60 percent of their operating and maintenance costs from farebox and parking revenues. The TSM alternative would require a 1995 operating subsidy of $5.8 million plus another $8.6 million a year in subsidy to cover capital costs. The rail alternative most like DartRAIL, by contrast, would require a 1995 operating subsidy of $7.8 million plus an annual capital subsidy of $65 million. Furthermore, none of the rail alternatives would be eligible for federal capital grants, the consultants reported, because they attracted so few additional riders and cost so much more than the TSM alternative.[14]

The consultants also examined DartRAIL's proposed TSD as a funding mechanism and found it wanting. The consultants argued that DartRAIL's revenue estimates assumed extremely optimistic forecasts of future office and retail construction and of real property value increases in the corridor. The proposed DartRAIL TSD also covered a much larger geographic area than the consultants thought could be justified by the improved transit service.[15]

The Fairfax County Board of Supervisors accepted the consultants' recommendations and adopted a bus-based system incorporating the

13. Draft reports were circulating by March 1989; the final report was released much later. See KPMG Peat Marwick, *Dulles Airport Access Road Corridor Transit Alternatives: Technical Report*, report prepared for the County of Fairfax (June 1990).

14. The FTA had more applications for new rail grants than it could fund and had devised a cost-effectiveness index to screen rail proposals. No new rail project would be considered unless the added cost for each additional ride attracted was less than $6.00. On the FTA's cost-effectiveness index, even the cheapest Fairfax County rail alternative was $19.16 for each added ride. The alternative most similar to DartRAIL was the worst performer at $25.84 for each added ride.

15. Needless to say, DartRAIL had many objections to the consultants' findings; for an account of these see Gómez-Ibáñez and others, *Prospects for Private Rail Transit*, pp. 89–90.

features of the TSM option with some enhancements. This option would be far more cost effective than rail, the supervisors argued, and would retain the option of building a rail system at a later date by preserving the sites of future rail stations as park-and-ride lots for the express bus system. Unlike the rail alternatives, the TSM options would be eligible for an FTA grant to cover 50 percent of the capital cost ($36 million). The remaining local share would be modest enough that it could be funded without resorting to TSDs or other new local taxes.

DartRAIL was also encountering difficulties in organizing its proposed TSD district even as county officials were opting for the bus alternative. To appeal to the property owners, DartRAIL had included the expensive Tyson's Corner monorail and upgraded the main rail line to make it compatible with Metrorail's heavy rail system. Many property owners were still convinced that the benefits they would receive would not be worth the added taxes they would have to pay, however, so DartRAIL never came close to the 51 percent of property value consents needed to create a TSD. DartRAIL's problems were compounded by poor timing. The softening of the property market in 1988, together with the county's more aggressive position on zoning, had combined to make property owners less receptive to the added financial burden of a TSD.

The Third Plan

Reacting to those difficulties, DartRAIL brought out a third plan in late 1990, which attempted to circumvent the objections of the county and property owners. DartRAIL proposed a three-step transit development process, beginning with bus and ending with rail. The first step would be express bus service along the Dulles Airport Access Road using stations in the median, built so that most of the structures could be eventually reused as rail stations.[16] Later a rail line would be built in the median and shuttle service operated between the Dulles and West Falls Church stations. Finally, the rail line would be connected to the Metrorail system to permit through service. The final rail system differed from the previous DartRAIL plan in that the rail line would depart from the

16. As a result, DartRAIL's express buses would be run like a train, with passengers transferring from separate feeder bus lines at the bus stations above the median. In the county's TSM plan, by contrast, bus stations and park-and-ride lots were located on the side of the Dulles Access Road/Dulles Toll Road right-of-way with access ramps to permit buses to enter and exit the uncongested access road. The ramps would allow integrated feeder-express service on routes with sufficient demand, which would avoid the inconvenience of transferring at the bus station.

median to pass through Tysons Corner instead of serving the area with a monorail loop.

To reduce reliance on TSD revenues, the third plan proposed tapping three new revenue sources. First, the funds Fairfax County had budgeted for its own TSM plan ($36 million in the county's own funds plus the promised FTA matching grant of $36 million) would be used to help build the initial bus system. Second, the airport would make cash contributions to help defray the costs of upgrading the system to rail; such contributions were now more likely, DartRAIL argued, because Congress had recently authorized airports to collect a passenger facility charge of up to $3 per person. Finally, DartRAIL proposed that the surpluses the state was earning on the existing Dulles Toll Road, which showed the same right-of-way as the Dulles Airport Access road, be used to help finance costs during both the bus and rail phases of the project.

DartRAIL hoped to use the toll road surpluses as a source of political leverage as well as financial relief. A regional airport authority, the Metropolitan Washington Airports Authority (MWAA), had been created to take over the operation of Dulles and National Airports from the FAA in 1987. MWAA had stayed out of the DartRAIL debate until the county adopted its bus plan. Shortly thereafter, however, the MWAA board, at the urging of its planning committee, passed a resolution informing the Commonwealth of Virginia that it would not grant an easement for a proposed extension of the Dulles Toll Road unless the surpluses on the existing toll road "were dedicated to transportation needs, principally rail, in the Dulles Corridor." The airport had leased to the state the land for the original toll road at no cost, and the state was now earning surpluses on the road. The MWAA board thought it had a right to say how those surpluses should be used, and some members strongly supported a rail link as long as it did not require a significant financial contribution from the airport.

The legal authority over the existing state toll road and its surpluses rests with the Commonwealth Transportation Board (CTB). Responding to the new controversy, in September 1990 the CTB committed itself to the development of "a comprehensive, phased, multi-modal transportation program including rail service as its transportation objective for the Dulles corridor" that would be funded by Dulles Toll Road surpluses. A draft of the program was to be released by December 15, 1990, for public comment and hearings with the final program to be promulgated by July 1, 1991. DartRAIL released its new plan shortly after CTB's September announcement in the hope that the DartRAIL proposal would be a compromise acceptable to all.

While DartRAIL had been preoccupied with preparing its new proposal, however, the FTA had informed the county that the $36 million in federal grants reserved for its TSM project would be released for other purposes if the county did not complete its grant application and secure the local match by December 31, 1990. The county responded by placing a bond issue for the local match on the November 1990 ballot, which voters approved. In the second week of December, after DartRAIL announced its new plan but before CTB's December 15 draft plan release, the board of supervisors selected the final locations for four bus stations and parking lots between Tysons Corner and Herndon.[17] At the board meeting, county officials argued that these site choices did not conflict with the CTB's approach since the facilities could be converted to rail stops once a feasible rail plan was developed. But while the bus stops and parking lots were located in the general areas long proposed for rail stations, they were sited just outside the Dulles Airport Access Road/Dulles Toll Road right-of-way rather than in the median, which would require a long walk for any eventual rail passenger. The county's plan might reduce the cost of an eventual rail system somewhat, but not nearly as much as DartRAIL's proposed first-stage bus system would.

By the fall of 1991 it was clear that the MWAA support and the CTB resolution had not provided DartRAIL with as much leverage as it had hoped. DartRAIL, however, was still optimistic. The county's insistence on building its bus system represented a lost opportunity, DartRAIL conceded, but was far from a crippling blow, and most observers also expected that DartRAIL would come up with yet another plan.

The Orlando Maglev Demonstration Project

A proposal to build a thirteen-mile magnetic levitation (maglev) train to connect Orlando International Airport and a terminal on International Drive near Walt Disney World is our third and final case study.[18] The Orlando maglev proposal grew out of Florida's efforts to promote high-speed intercity rail transportation. In the late 1970s the then-governor of Florida, Rubin Askew, went on a trade mission to Japan during which he

17. Steve Bates, "Fairfax Picks Bus Stops for Dulles Access Strip," *Washington Post*, December 13, 1990, p. C5.

18. For a more detailed account of this case see Gómez-Ibáñez and others, *Prospects for Rail Transit*, pp. 111–63.

rode the high-speed bullet train from Tokyo to Osaka. The governor was impressed and suggested that Florida needed this kind of technology. He assigned a young Japanese-born assistant, Sam Tabuchi, to explore the idea with officials of the Japanese National Railway (JNR), but a full investigation of a Florida high-speed rail system failed to materialize. Several years later Tabuchi returned to Japan accompanying Governor Bob Graham; Graham shared his predecessor's enthusiasm for high-speed rail but took it further, establishing in 1982 a High-Speed Rail Committee (HSRC) to investigate the feasibility of building a high-speed rail system in Florida.

Early Planning

The HSRC quickly discovered that high-speed intercity rail systems would probably require public subsidies because of Florida's relatively low-population densities and the modest travel volumes in its major corridors. Rail supporters convinced the HSRC, however, that the profits from land development stimulated by the rail system might be sufficient to cross subsidize rail losses, so that a new rail system could be built at no direct government expense. Accordingly, the HSRC recommended a program, passed by the legislature as the High-Speed Rail Act of 1984, that offered four key incentives for high-speed rail development: use (through Florida's Department of Transportation) of the power of eminent domain to appropriate private property for the right-of-way; use of certain publicly owned corridors for right-of-way; tax-exempt financing to be secured by system revenues;[19] and a means of capturing any added real estate development values created near stations and along the corridor.[20]

By the mid-1980s Sam Tabuchi had left state government to work as JNR's Florida coordinator, investigating the potential for submitting an application for the high-speed rail franchise. Tabuchi quickly realized that one of the few segments with adequate traffic to justify a high-speed rail system was the corridor between Orlando International Airport and Walt Disney World (WDW). While the thirteen-mile airport-WDW seg-

19. The amount of tax-exempt financing would be limited by the state's overall annual issuances of $625 million. The potential franchisee was also encouraged to use other innovative forms of finance such as tax increment districts and foreign credits.

20. Real estate values could be captured in two ways. First, the rail franchise could build ancillary developments around stations without going through Florida's often-burdensome local land use permitting process. Second, the rail franchises could establish a special benefit assessment district to assess nearby developers for the value created. See Gómez- Ibáñez and others, *Prospects for Private Rail*, pp. 114–15.

ment alone could not subsidize the extensive intercity rail system Florida was hoping for, it offered an attractive opportunity for a commercial demonstration of JNR's maglev technology, and one in which maximum operating speed could be achieved. In addition, the segment was short enough to keep overall guideway costs (of $10 million–20 million a mile) within reasonable range.

With a preliminary expression of interest from Disney, Tabuchi approached JNR for funding to develop an Orlando demonstration. JNR put him in touch with the Forum for Urban Development, a group of Japanese urban planners, industrial corporate officers, and politicians interested in promoting innovative solutions to urban development problems. The forum brought together several Japanese banking and industrial interests to help support the effort, including C. Itoh, a member of the forum and one of Japan's largest trading companies, and the Dai-Ichi Kangyo Bank, the world's largest bank (measured by assets). In the meantime, however, JNR was privatized (in 1987) and its research program, including work on maglev systems, was cut back and focused on domestic Japanese applications. Prospects that JNR would be able to support the export of its technology to the United States were thus greatly reduced, and the forum was forced to consider alternative systems.

Such an alternative had been developed by a consortium of West German firms known as Transrapid. Transrapid operated a twenty-mile test track in Elmsland, West Germany, where its train had achieved speeds of over 250 miles an hour. The forum decided that it would proceed with the Orlando demonstration if Transrapid could be brought in as a partner. For its part, Transrapid was anxious to realize a commercial demonstration. The West German government had invested over $1 billion in maglev research and developed the world's most advanced system, but without a revenue-generating demonstration, funding might be directed toward other activities. With an international partnership capable of delivering the technology as well as the financing in place, Tabuchi turned to securing support from public and private interests in Florida.

The Maglev Demonstration Act

Tabuchi and his partners drafted a Maglev Demonstration Act, modeled after the High-Speed Rail Demonstration Act but with modifications to broaden support. The new bill incorporated two key provisions of the High-Speed Rail Act: a one-stop permitting process and the use of eminent domain to help acquire right-of-way and station sites. Like the previous act, the Maglev Demonstration Act also prohibited direct government subsidies, but it went even further in eliminating the authorization for the use of tax-exempt bonds and special assessment districts or other devices to capture profits from ancillary real estate developments. Moreover, the new bill specified that the maglev demonstration had to be confined to a single county to overcome opposition from the firms competing for the statewide high-speed rail franchise. On the surface the act seemed to offer an unbeatable combination: a high-tech transportation system and tourist attraction for central Florida that would not cost taxpayers a penny or threaten neighboring development interests. Moreover, maglev demonstration applicants would also have to pay a half million dollar filing fee to cover state administrative costs.

The Maglev Demonstration Act outlined a three-step procedure for obtaining a franchise. First, the Florida High-Speed Rail Transportation Commission, established by the High-Speed Rail Act, would solicit and review applications. After a public hearing, the commission would recommend whether applicants should be certified, based on whether the proposal was financially feasible and that no government funds were needed or requested. Second, an administrative hearing would be held under the auspices of the State Department of Administration for the purposes of airing disputes unresolved by the commission and making recommendations on certification, including conditions of the project. Finally, the recommendations of the commission and the administrative hearing officer would be forwarded to the governor and his cabinet for a final decision on certification.

The Maglev Demonstration Act quickly won approval at the start of the 1987 legislative session, and Tabuchi and his Japanese backers incorporated Maglev Transit, Inc. (MTI) in Florida, with Tabuchi as president, to file an application for the maglev franchise. Members of the Forum for Urban Development, who had already spent several million dollars in supporting Tabuchi's efforts and acquiring right-of-way options, supplied the financial backing for the new corporation.[21]

MTI's First Proposal: Airport to EPCOT

On March 2, 1989, MTI filed an application with Florida's High-Speed Rail Transportation Commission for certification of a maglev

21. The forum's earlier efforts were undertaken by Florida Environmental Resources, another Florida corporation.

demonstration. The proposed twenty-mile route would run from a station adjacent to the new southern terminal of Orlando International Airport to a terminal northwest of Disney's EPCOT center. The trip would take seven minutes and operate with fifteen-minute headways.

Since its alignment made extensive use of existing public airport and highway right-of-way, MTI did not expect to buy a significant amount of land.[22] And some of its own environmental impacts, such as train noise, would be masked by the facilities with which it "cohabited." MTI expected to recover all its costs from farebox revenues, including its projected investment of $520 million (exclusive of construction loan interest). Each mile of fully equipped guideway would cost over $11 million. Trains would cost another $120 million, and stations and operations and maintenance facilities $116 million. Annual operating costs would be approximately $50 million. Approximately half the riders were expected to be transferring between the airport and Disney or nearby hotels. The remainder would be "attraction" riders—passengers who would pay for the thrill of riding on a high-speed and technologically advanced train.

The High-Speed Rail Transportation Commission scheduled public hearings for June 1989.[23] Opposition had been developing even before MTI submitted its application, however, and the opponents came out in force at the June hearings. Indeed, even as it was developing its proposal, MTI had received complaints from local political leaders and business interests that its system would primarily benefit Disney.

These concerns were felt most acutely by the competing theme parks and businesses along International Drive. International Drive is located along the eastern edge of WDW and thus is an ideal site for attractions and hotels designed to divert tourists heading to Disney as well as attracting a market of their own. Sea World anchors the southern end of the drive, while Universal Studio's theme park is located toward the northern end. In the middle is Wet 'N Wild, a water park with slides, chutes, and pools. Motels and high-rise hotels, providing almost 14,000 rooms, and many different restaurants and shops line the corridor between these attractions.

22. Almost 44 percent of its route would following existing public rights-of-way. MTI estimated land acquisition costs for other segments at $10 million but did not include costs for leasing public rights-of-way.

23. Other groups had reportedly considered filing an application, including a consortium that proposed to use the high-speed surface transportation maglev system that was being developed by Japan Air Lines.

International Drive property owners sought something that would sour the MTI deal for Disney or at least make the maglev line work in International Drive's interests as well. They found it in a clause of the Maglev Demonstration Act, which required public access to transit systems. International Drive interests had long hoped to develop a light rail line or people mover along International Drive. They now argued that their light rail proposal should be revived and coordinated with MTI's maglev line by conditioning MTI's franchise on the requirement that a light rail station could be built adjacent to the maglev station on Disney property. By reviving the light rail system, International Drive interests cast themselves not as the spoiler of MTI's deal but as champions of a wider community interest. To counter this public access concern, moreover, MTI had already indicated in its application that it would study the feasibility of constructing a spur and station to International Drive once its main line was up and running.

The commission, after hearings, came down on the side of International Drive, in a decision delivered on June 29, 1989. MTI was directed to provide agreements that the Airport Authority and Disney would insure public access by rail or other means of ground transport and that it would cooperate in conducting a feasibility study to determine the type, location, and financing for any means of ground transportation that complied with the accessibility requirement of the act.

WDW executives were not happy about these conditions. Although not mandating a specific solution (for example, a light rail transit system), they effectively would bind Disney to negotiate with its Orlando area competitors over a transit system. Maglev as an additional attraction was one thing; but as an offer that might affect Disney's control over its own property, it was quite another.

In this connection, Orange County officials had been negotiating for years to get WDW to pay for the external impacts of its operations. In an apparent gesture of accommodation, Disney agreed to a multimillion dollar payment to the county while the county, in return, agreed not to seek further compensation from WDW or to engage in other legal actions that might affect WDW's development for ten years. But before the county could savor its victory, WDW announced that it would be building a 1.5 million square foot mega-mall in the southeast corner of its property at the Osceola/Orange County border. In one off-the-record conversation, an area attraction owner observed, "Disney is doing it to us again. They realize that one of the biggest tourist attractions off their property is outlet stores, and now that others have discovered and developed that market they're trying to capture the tourist-shoppers for themselves." What Disney had in mind, however, was not outlet stores but an up-scale regional mall containing the likes of Neiman Marcus and Harrods. The new development would include 5 million to 8 million square feet of commercial office space, thousands of hotel rooms, and hundreds of housing units.

In October 1989, WDW suggested that its new mall site could give MTI a solution to the conditions now attached to its application: run the maglev there, where there was plenty of land to meet the accessibility requirement. Indeed, WDW would provide unrestricted public access. Presumably this offer would be attractive to some of MTI's Japanese investors who were still looking for land development opportunities at WDW, but, since the new site was about seven miles away from EPCOT, it was decidedly less advantageous as a site to generate many attraction riders. WDW assured MTI that its internal transit system would be extended to the site. The new site was also farther from the airport and, at over U.S. $11 million a mile for fully equipped guideway, ridership would have to be far higher than previous projections for the EPCOT site if all costs were to be covered. A WDW terminal had always seemed like the way to make the demonstration work financially, but clearly not any destination at Disney would do. MTI needed time to regroup. If a WDW terminal site was no longer realistic, then an alternative would have to be found and an amended application filed. But was there another site capable of generating adequate ridership?

The Second Proposal: Airport to the Grand Terminal

MTI, International Drive property owners, and city and county officials developed a new scheme and on March 9, 1990, MTI filed an amended application with the High-Speed Rail Transportation Commission. The new proposal deleted the EPCOT station and truncated the line at a new Grand Terminal on the southern extension of International Drive. The new route would be slightly over thirteen miles, following the original alignment up to the new station site. The Grand Terminal would consist of the maglev station, parking structures, and a light rail transit facility. In phase one, a complex consisting of a 1,500-room hotel and a 300-room executive conference center would be built immediately north of the terminal. In phase two another thirty acres to the north would be developed with 1,500 hotel rooms along with half a million square feet of office space. The Grand Terminal would be a transit hub, offering rental car service, vans, buses, and a light rail connector to Sea World, Wet 'N Wild, the Convention Center, and Universal Studios. The terminal also would offer an extensive exhibit educating guests about the physics of magnetic levitation and promoting its use in transit. The developers of the Grand Terminal would be expected to build the station at no cost to MTI.

With a route approximately a third shorter than its original proposal, MTI estimated that capital costs, exclusive of the Grand Terminal Station and interest during construction, would be about 14 percent lower. Major savings—almost $20 million—were projected for land acquisition and construction equipment.[24] Operating and maintenance costs were expected to be about $44 million, 7 percent less than in the original proposal, but the internal rate of return on equity dropped from 20.74 percent to 16.6 percent.[25] A sensitivity analysis included in MTI's amended application concluded that in a worst-case scenario, return on equity could fall to –0.5 percent.[26] As with its original application, MTI projected rider's fares as its principal source of revenues.

New hearings before the High-Speed Rail Commission on the amended application began in April 1990. As with MTI's earlier proposal, the commission retained Price Waterhouse to perform an independent evaluation of MTI's financial plan.[27] It concluded that the logic, assumptions, and projections of the plan were internally consistent. However, it also found the plan especially sensitive to changes in revenue and

24. Since most of the additional route to EPCOT would have been on a WDW right-of-way (presumably provided at no cost), it is not clear where these savings are derived. The new cost estimates for guideways are consistent with those suggested by the Federal Railroad Administration (FRA). See *Assessment of the Potential for Magnetic Levitation Transportation Systems in the United States*, Report to Congress by the Federal Railroad Administration (Washington, June 1990). Since the FRA had consulted with Transrapid in developing its estimates, it would appear that MTI's original estimates were low or that Transrapid had recalculated guideway costs in the intervening period between MTI's applications.

25. At a hearing of the High-Speed Rail Transportation Commission on April 16, 1990, Tabuchi suggested that although the financial analysis of the amended application excludes income from real estate development at the Grand Terminal, investors with combined interests in Maglev and its associated real estate would realize an IRR in the 20 percent range.

26. The worst-case scenario assumed that interest on debt would be 2 percentage points higher, construction costs 10 percent higher, and ridership 20 percent lower than the most likely scenario.

27. Price Waterhouse, *Financial Evaluation of Maglev Transit Inc.'s Amended Application to the Florida High-Speed Rail Commission* (April 1990).

operating costs. Price Waterhouse also wondered how willing the investors in the Grand Terminal's ancillary real estate developments would be to help MTI. Despite the many reservations raised in the Price Waterhouse report, the commission again expressed confidence in the reasonableness of MTI's financial plan and recommended certification. Hearings before an administrative officer were scheduled to take place within ninety days, with the matter then to be placed before the governor and the cabinet.

After all of its careful negotiations MTI expected to make its case at the administrative hearings with little opposition from its former opponents. For the most part this assumption was correct. State and local governments, International Drive interests, and Disney raised no significant objections; but not the Greater Orlando Aviation Authority (GOAA), which operated the Orlando Airport. GOAA was primarily concerned with MTI's ridership estimates, suggesting that they were inflated and that the system could be in serious financial trouble once it began operations. The airport believed MTI had never proven financial feasibility, and that if certified, the airport and local governments would somehow be left with financial liabilities for keeping the system running, or for removing it, if it went bankrupt.

An order on the magnetic levitation demonstration project was issued by the hearing officer in February 1991. The hearing officer agreed with GOAA that MTI had underestimated several items in its cost estimates and that MTI's ridership projections were flawed and revenues consequently might be lower than projected. Nevertheless, he concluded that his reservations were "not sufficient to establish that public funds would be needed for the proposed project or that the proposed project cannot be constructed and operated without public funds or government subsidies."[28] With this finding established, the governor and cabinet, meeting on June 12, 1991, gave final approval to the maglev demonstration.

Winning certification was an impressive achievement, but it had taken three years to devise an acceptable plan, much longer than MTI had expected, and the approved scheme was more complex and risky than MTI's original design. As of 1992, MTI still had not convinced investors that its $600 million demonstration is financially viable. State certification was not an endorsement that the project is investment grade; the High-Speed Rail Commission had argued that the 1987 Maglev Demon-

28. State of Florida, Division of Administrative Hearings, case number 89- 3563, in *Re: Magnetic Levitation Demonstration Project, Maglev Transit, Inc.* (February 22, 1991), p. 213.

stration Act only requires that the project not be a potential financial burden on government. As a demonstration, the maglev line might indeed prove unprofitable, but that would be acceptable as long as private investors, and not the public, bear the risks.

The Problems of Private Rail Proposals

These three case studies suggest that difficulties with privatizing rail transit arise from three sources. First, rail transit systems are so expensive to build, operate, and maintain that recovering costs through farebox revenues alone is usually very difficult. Construction costs, as well as operating and maintenance costs, are substantial for rail systems. Furthermore, ridership demand is often slow to mature, and the characteristics of rail systems make it difficult to adjust costs and operations incrementally as demand grows. As a consequence, operating and maintenance costs per passenger are high in early years which, combined with heavy debt service, can result in a protracted period of negative cash flow. Even assuming ridership reaches its projected peak after several years of operation, an assumption not always borne out, farebox revenues are often inadequate to cover operating and maintenance costs, much less capital costs.

Second, in an effort to cover costs, private sector providers seek out other sources of revenue, typically from land development, to subsidize the rail offering. Indeed, often the private guideway offering is effectively a secondary activity within a wider set of investments. From the start, developers of DartRAIL and the World Trade Center monorail looked to real estate developments to subsidize system costs, and both eventually called for the establishment of special transportation assessment districts to enable their financing. For Orlando's maglev demonstration, even though ridership projections were exceptionally high, feasibility would require that the technology provider (Transrapid) absorb capital cost overruns (through a fixed-price contract) and the Grand Terminal developers subsidize station construction costs. Private rail developers also frequently seek direct or indirect public subsidies. As a result, these projects rarely turn out to be pure transit offerings or purely private.

Third, in rail transit privatization the search for ancillary revenues results in a widening circle of interested parties and project participants, both private and public, inevitably leading to greater complexity in forming and sustaining coalitions and partnerships. Ultimately, these three conditions combine to produce a powerful political-economic dynamic that stalls the projects. This dynamic is a problem with almost all infrastructure privatizations but is most troublesome for rail transit.

Rail transit systems do have advantages in attracting external support, however, since they are fixed and therefore better able to focus benefits. If most of the benefits of a rail system are realized within a quarter mile radius of its stations, then fixing station sites and controlling adjacent land maximizes value capture. Such focus makes it easier to identify and solicit financial support for the system from the property owners near the station areas. DartRAIL could add a loop at Tyson's Corner or alter station locations, not simply to serve the greatest concentration of employment destinations but also to direct benefits to the landowners whose support they needed. In Boston, the monorail providers expected that other developers in the area would financially support the system if stations were located nearby.

The location of rail stations, however, not only makes it clear who is likely to benefit but also who may be placed at a competitive disadvantage. Disagreement about these distributional questions created serious problems in all three cases. In Orlando the original choice of EPCOT as a terminal site announced who would benefit and simultaneously provided property owners along International Drive with a reason for opposing it. That the project was offered as a publicly sanctioned demonstration gave broad standing to opponents and a platform from which to try to delay the project's certification. Similarly, developers of Boston's World Trade Center portrayed the monorail as mitigating the transit impacts of their real estate development, but the neighboring developers of Fan Pier saw it as complicating their efforts to get development approvals while simultaneously conferring a potential competitive advantage on a rival.

All this suggests that rail transit offerings may not be the most appropriate mode for infrastructure privatization initiatives. Highway bridges and tunnels, for example, may be much easier to implement because the chances of user fees meeting all financial requirements are greater, thereby alleviating the need to capture external or other benefits. Almost by definition, financial self-sufficiency is an important aid to privatization, not only by meeting the most important of market tests but also by providing the wherewithal to avoid the complications of political entanglements.

Chapter 14. The Political Economy of Profitability and Pricing

with ARNOLD M. HOWITT

THE EXPERIENCES WITH AIRPORTS and urban rail transit suggest that privatization is politically more acceptable when the project is neither highly profitable nor highly unprofitable. But how well does this political perspective mesh with economic or efficiency considerations? Do efficiency considerations also dictate that transportation services, whether public or private, recover their costs from prices charged users?

In this chapter we contrast the economic and political perspectives on the profitability and pricing of privatized transportation services, drawing on experience with private toll roads and buses as well as with airports and urban rail transit. We contend that the economic and political prescriptions for pricing and profitability are remarkably similar when *new* private services or facilities are provided but differ when *existing* facilities or services are privatized.

Profitability and Economic Efficiency

Economists distinguish two types of efficiency: technical and allocative.[1] Technical efficiency concerns how effectively a firm or an industry uses its resources. A firm is technically efficient if it produces its output by

1. Economists define profits differently from accountants or the general public because they treat returns to equity and entrepreneurial effort differently. Accountants typically define any and all return to equity and entrepreneurial effort, left over after meeting all other costs, as profits. An economist considers a competitive return on investment (the return required to keep capital committed to the activity) part of normal costs. Without that return, capital would be withdrawn as opportunities became available (for example, physical exhaustion of the plant). Hence, for economists, these normal competitive returns represent legitimate costs of the enterprise. Economic profits only accrue when the firm earns an above normal or excess return on its investments, that is, more than needed to keep capital committed. To avoid confusion, we will call economic profits surpluses or excess profits or economic rents, and economic losses simply losses or, where they are borne by the public sector, subsidies. It should be understood that an economic loss can occur even when accounting profits are somewhat positive (if total revenues are insufficient to keep capital committed).

using the least-cost production techniques and input combinations. Allocative efficiency concerns whether the firm or industry is producing the appropriate output given society's competing needs for other goods and services. Allocative efficiency implies rules for both pricing and investment in new capacity.

Technical efficiency does not require a firm to recover its costs from the prices it charges users. What seems more important is that the rules for any public subsidies are structured so as not to distort the firm's choices in production technology, and that the firm is motivated by the possibility of earning an excess (or above normal) profit if it can reduce its production costs. In this context, privatization is often attractive as a means of creating the profit motive for technical efficiency, as the British and U.S. experiences with competitively contracting out subsidized transport services demonstrate. Nevertheless, technical efficiency seems more easily achieved when the firm is not reliant on public subsidies. Administering a subsidy program without undermining incentives for technical efficiency can be difficult, as the subsidy programs for public bus companies in many countries and private toll roads in Spain and France illustrate. Without subsidies, the rules for achieving technical efficiency are simpler.

For allocative efficiency, economists usually recommend that prices be set at the short-run marginal costs (SRMC) of accommodating another user. Basically, pricing at marginal costs encourages consumers to use a good or service only when the benefit received is worth at least as much as the cost of providing it. The short run is the economists' term for the period in which the basic capacity of the transportation facility or service cannot be adjusted, say, by adding an additional traffic lane on a highway, building a new terminal or runway for an airport, or adding to a fleet of urban buses. For highways, for example, the short-run marginal cost (SRMC) does not include costs of acquiring right-of-way, grading, and constructing the road surface since such costs are fixed or sunk in the short run. The major components of SRMC are the traffic delays or congestion that an additional motorist imposes on other highway users and the wear and tear on pavement caused by passage of an additional vehicle. Prices based on SRMC would vary by type of user, time of day, and location because congestion, pavement damage, and other such short-run costs will so vary.

The appropriate level of investment allocated to a transportation activity is determined by comparing the SRMC with the long-run marginal cost (LRMC). The LRMC incorporates the cost of accommodating additional traffic by physically expanding the facility. For highways, LRMC includes investments in land, grading, structures, and so on, as well as operating and maintenance expenses (but not congestion costs). The LRMC is usually higher for traffic in the peak period and direction than for the off peak because facility expansions are primarily needed to accommodate additional peak traffic.

Investment is optimal if the SRMC, and thus prices, equals the LRMC.[2] On highways, for example, additional traffic on a road can be accommodated in two ways: by tolerating slower speeds and additional congestion and wear and tear on the facility, thereby incurring SRMC, or by physically expanding the facility so that additional highway traffic does not slow down highway speeds and add to congestion, thereby incurring LRMC. If SRMC exceeds LRMC, the road is too congested and small because it would be cheaper to accommodate more traffic by building more capacity than by tolerating more congestion. Conversely, if LRMC exceeds SRMC, the road is too big because it would be cheaper to tolerate more congestion than add to capacity.[3] When LRMC and SRMC are equal, and user charges are set at SRMC, the levels of investment and congestion should be just right.

These pricing and investment rules imply three simple conditions under which a transportation firm will just recover its total investment and operating costs from efficient prices.[4] First, there must be no significant external benefits or costs to building or using the transportation service beyond those that accrue to the builder or the facility's users. Such external costs might include noise or air pollution imposed on nonusers, for example, while external benefits might be congestion relief on competing modes or facilities. Absent such externalities, all the components of LRMC are paid by the facility supplier.[5]

2. See, for example, Theodore E. Keeler and Kenneth A. Small, "Optimal Peak-Load Pricing, Investment, and Service Levels on Urban Expressways," *Journal of Political Economy*, vol. 85 (February 1977), pp. 1–25; and J. Brechman and D. Pines, "Financing Road Capacity and Returns to Scale under Marginal Cost Pricing," *Journal of Transport Economics and Policy*, vol. 25 (May 1991), pp. 177–80.

3. Strictly speaking, this situation assumes that the costs of expanding and reducing capacity are the same. This may not be true for many transportation activities, at least short of very long-term adjustments in right-of-way and other very fixed investments.

4. Other combinations of conditions can also lead to a situation in which efficient prices just recover a firm's costs. Modest external costs might be just offset by economies of scale in the firm's own costs, for example, since the former dictates efficient prices in excess of the firm's costs while the latter dictates prices below the firm's costs.

5. SRMC includes costs to the users as well in the form of the added congestion on the firm's facility created by increased traffic.

Second, investment must be optimal, or near optimal, for current demand so that the efficient price equals LRMC as well as SRMC. Investment might not be optimal, for example, if demand was declining or expanding so rapidly that capacity could not be adjusted to keep pace. Alternatively, if there are economies to adding capacity in fairly large increments, it may pay, when anticipating future growth, to add more capacity than immediately needed.

Finally, there should be constant returns to scale in the firm's long-run costs of building and operating the facility.[6] Constant returns imply that pricing at LRMC will just recover the firm's long-run total costs. If there are increasing or decreasing returns to scale, however, then user charges set at LRMC will yield revenues that fall short of, or exceed, total costs, respectively.

Two of these three conditions are often met in the real world, or at least not violated so much as to cause a serious conflict between profitability and efficient pricing. Many transportation modes exhibit constant, or near constant, economies of scale in long-run costs, especially highways and buses.[7] Urban rail transit is an important exception in that most studies show increasing returns to scale. These economies are strongest at traffic volumes where rail is less cost effective than other modes, especially buses or automobiles on highways, however, so that such economies are important mainly in places where a rail line should not be built in the first place.[8]

The problems caused by nonoptimal capacity are often less serious than they might seem, especially in modes such as buses using public streets. Bus capital comes in small increments (a single bus) and has a relatively short life, so capacity can be quickly adjusted. Even for capital-intensive and long-lived infrastructure—such as roads, rail transit, or airports—the problem may not be severe. Capacity can often be increased

6. More accurately, there should be constant returns to traffic density rather than scale. Traffic density is the volume of traffic on a given route or network of constant length. Increases in scale can include expansions of the length of the route or network as well as increases in traffic density. The distinction between density and scale is rarely used or relevant in nontransportation industries. Although the term *scale* is too inclusive, we use it here because our readers are more likely to be familiar with it.

7. See Marvin Kraus, "Indivisibilities, Economies of Scale, and Optimal Subsidy Policy for Freeways," *Land Economics*, vol. 57 (February 1981), pp. 115–21; Marvin Kraus, "Scale Economic Analysis for an Urban Highway Networks," *Journal of Urban Economics*, vol. 9 (January 1981), pp. 1–22.

8. See, for example, John R. Meyer, John F. Kain, and Martin F. Wohl, *The Urban Transportation Problem* (Harvard University Press, 1965), pp. 299–306.

incrementally. Highway capacity can be increased gradually by adding lanes only at key bottlenecks or by slowly eliminating serious grades or dangerous curves, for example, while airport capacity may be increased one terminal or runway at a time. Where gradual expansion is not cost effective, moreover, prices set at SRMC will still fluctuate around LRMC, typically being lower in the years soon after capacity investments are made and higher in the years immediately before that time. Thus costs can often be recovered over the life of the investment even if prices are set at SRMC, with the early losses being offset by later profits.

Externalities can be more of a problem, although they are as likely to dictate excess profits as losses. The most common and significant externality leading to loss of profit is congestion relief on competing modes and facilities that are priced below SRMC. If buses or rail transit reduce automobile traffic and congestion on underpriced urban roads, or a new toll road relieves congestion on a competing parallel free road, for example, then reducing the bus or rail fare or the highway toll below the firm's LRMC to encourage traffic diversion may be desirable. Often the fare or toll has relatively little effect on diversion, however, so the social benefits from reducing prices below costs is small. This proved true of British buses, for example. Increases in bus prices after privatization had little discernible effect on auto traffic. For toll roads the record is mixed, with tolls causing little effect on parallel routes in France, for example, but more serious problems in Spain and Mexico, where toll rates are higher.

Noise, air pollution, or community disruption from building and operating a transportation facility or service are externalities that can justify prices above the transportation firm's costs. Often these external costs are relatively small compared with the supplier's costs, so the necessary markup is not significant.[9] In developed countries, the transportation provider also may be required to mitigate the damage, by building noise barriers or landscaping, for example, so that external costs are effectively built in to the firm's accounts.

An important exception occurs, however, when environmental and community opposition to highway, airport, or other infrastructure expansion is so strong that it is in effect politically impossible for developers to expand capacity. This problem usually occurs in built-up areas where congestion, and thus prices based on SRMC, are likely to be very high.

9. Theodore E. Keeler and Kenneth A. Small, *Automobile Costs and Final Intermodal Cost Comparisons*, part 3 of *The Full Costs of Urban Transport*, ed. T. E. Veeler, L. A. Merewitz, and P. M. J. Fisher (University of California at Berkeley, Institute of Urban and Regional Development, 1975).

The proceeds from the user charges could be used to repay the original costs of constructing the existing roadway or airport or of maintaining the facility. But prices based on SRMC are likely to yield substantial amounts above and beyond such costs when demand is strong and expansion of capacity is constrained.

Other practical considerations strengthen the argument that efficient pricing and investment are usually consistent with a mandate that transportation suppliers recover their costs from user charges. Substantial practical problems can occur in administering a strict regime of SRMC pricing, for example. A good deal of information is required to set the correct prices for different users imposing different costs on the system (by time of day, vehicle axle weights, congestion level). The costs of administering and collecting these different tariffs could be substantial and possibly outweigh the allocative efficiency benefits. Such complex fare structures, especially if somewhat unpredictable, may also induce mistaken investments downstream by some users of the facility. If a strict regime of marginal cost pricing requires government subsidy, moreover, the allocative inefficiencies caused by increasing taxes to finance the subsidy must be considered. As noted earlier (in chapter 6), raising taxes by $1 may create losses greater than $1 elsewhere in the economy because of tax-induced distortions and tax administration and collection costs.[10]

In some cases, moreover, the choice may be between a financially self-supporting new private service or facility and no new facility or service at all. Limits on public taxes or debt may make a private and self-supporting concession the only option for improving services, as seems to have been true of Mexico's toll roads and Morocco's buses. When the new service or facility is badly needed, the allocative efficiency gains may be larger from allowing it to be imperfectly priced, and thus financially self-supporting, rather than by not providing it at all.

In short, in some situations economic efficiency is inconsistent with supplier cost recovery or profitability. If external benefits or costs are unusually large, or it is impossible to expand capacity to accommodate growing demand, an efficient operation may incur losses or earn excess profits. Most of the time, however, efficient user charges should be approximately enough to recover the supplier's costs.

10. Edgar Browning, "The Marginal Cost Public Funds," *Journal of Political Economy*, vol. 84 (April 1976), pp. 283–98; and Dale W. Jorgenson and Kun-Young Yun, "The Excess Burden of Taxation in the United States," *Journal of Accounting, Auditing, and Finance*, vol. 6 (Fall 1991), pp. 487–509.

The Politics of Surpluses and Pricing

Excess profits, or surpluses, if they should arise, would seem at first glance to have many attractions for governments interested in privatization. After all, the goal of privatization is often to convert a money-losing public enterprise to profitability or to privately finance a facility the government cannot afford to build. And where surpluses are earned, the government may be able to capture part of them for other public purposes. Surpluses can create problems, however, depending on where they come from, how they are used, and how large they are.

The Sources, Uses, and Size of Surpluses

In privatization the source of any surpluses and whether they come from technical efficiency gains or transfers from other parties, such as labor or users, are important considerations. Pure technical efficiency gains are the least controversial. While demand for labor and other inputs may decline by building or operating a facility through the use of a lower-cost technology or input mix, the input market is usually so large that the decline will hardly be noticed. The cost-saving designs of the private French autoroute company, Cofiroute, for example, had a negligible effect on the total market for construction workers or materials in France. If efficiency gains are large enough, moreover, prices may be reduced or service improved to the point at which output increases and more inputs are employed. This happened in Britain, for example, after the bus reforms. Total operating employment increased despite dramatic improvements in labor productivity.

To the extent that surpluses are created by transfers, however, the losers are likely to be antagonized. The reductions in some drivers' wage rates that accompanied British bus deregulation were controversial, although their effects were softened by the increase in total employment and the fact that lower wages mainly applied to new hires. An even more common and controversial form of transfer is any increase in the prices charged users. While bus privatization in developing countries often occurs without price increases, highway privatization often means the imposition of tolls where free or tax-financed roads were formerly the norm.

The use of the surpluses is also potentially controversial. Surpluses retained by the investors or owners create the most potential for political unease. The possibility of earning surpluses is, of course, an essential motivator for the new investments or cost reductions sought from privatization. But the public may not appreciate these long-term benefits of occasional surpluses, the risks the successful investors and entrepreneurs originally took, or that some investors may have lost money in failed proposals or ventures. Moreover, surpluses may be viewed, often legitimately, as a sign that a firm is exploiting monopoly power. The public is thus likely to see surpluses as evidence that government has not done its job in protecting against market abuses.

Controversy is likely even if the government captures some of the surpluses for public purposes, say, through the sale price of a public enterprise or a profit-sharing agreement in the concession contract. While some citizen groups will benefit from surpluses, others may feel they have a claim and be disappointed. Certainly, any facility users contributing to the surplus through higher fees are likely to feel maligned.

These problems are more difficult the larger and more visible the surpluses. A small surplus is likely to pass unnoticed, but a large one probably will not. Hiding the surpluses may therefore be attractive. This is the motivation, for example, apparently underlying the cross subsidies in the French autoroute system; few motorists on France's heavily traveled north-south autoroutes probably realize that they are generating substantial profits which are being used to finance construction of more lightly traveled segments.

Demand Management versus Capacity Financing

The views of users of the facility or service are important since they often perceive themselves as the biggest source of any profits, surplus or not. Higher tolls or surpluses are more acceptable to users if they receive something in return. The most obvious compensation is improved service or increased capacity. In this respect, the general public and economists are attracted to pricing for different reasons, reflecting two distinct roles that pricing can play. First, pricing can encourage more efficient use of existing transportation facilities by managing or dampening demand, and second, pricing can finance improvements in capacity or service.

Economists tend to emphasize pricing's effect on demand. Demand management is the motivation for pricing at SRMC, and the reason economists are often concerned that prices should vary with levels of congestion. To be sure, prices also help signal whether or not investment is needed through comparisons of SRMC with LRMC. But, as already seen, the receipts from SRMC pricing equal investment and operating costs only under certain conditions.

The general public, by contrast, seems attracted to pricing by its potential to finance new capacity and services rather than its power to dampen demand. In virtually all cases where tolls have been imposed in the United States, for example, they were initially sold to the public as a means to finance the construction of a new highway, bridge, or tunnel that otherwise would not be immediately available. Tolls are often lifted once the initial construction costs are paid. Where tolls are continued, moreover, they are usually rationalized by using the proceeds to finance further improvements to the tolled facility or network or for reliever facilities. Not surprisingly, therefore, the tolls charged by public authorities usually do not vary by time of day or congestion, since revenue generation and capacity expansion, not demand management, are the primary motives.

Private companies seem more willing than users or the general public to employ pricing for demand management, largely because it is often consistent with maximizing profits or needed to attract requisite capital. The two proposed congestion-relieving private toll roads in California would use time-of-day pricing because owners know that motorists will pay more to drive on their facility when the alternative routes are congested (and on at least one of these roads, every possible source of revenue may be needed to earn an adequate return on capital). Where private companies are service users rather than facility providers, however, they tend to adopt the users' perspective. When selling their own services, U.S. airlines have created sophisticated systems of "yield management" to vary fares with demand and passengers' willingness to pay. But many airlines have traditionally opposed airports' proposals to vary landing fees by time of day or to charge any more than needed to recover the historic costs of airport investments.

Privatizing Existing versus New Facilities or Services

Given their emphasis on demand management, economists tend to view privatization and price increases on an existing facility, without improvements, far more favorably than the general public. If demand is growing but the capacity of a highway or an airport is fixed, for example, additional motorists or passengers can impose significant delays on others. In such congested situations, imposing prices that vary by time of day will force users to consider alternative facilities, modes, and travel times and reduce overall congestion. A private operator will have strong economic incentives to impose such prices, since doing so will also increase profits. The only potential drawback to privatization, from this perspective, is that the private operator might price above SRMC either because it enjoys some monopoly power or because the other competing facilities are underpriced and overcongested.

For the users, however, the newly imposed prices are a burden. If the revenue is retained by the private operator or captured by the public sector for other purposes, some users may gain from the imposition of prices, but many will lose. The principal winners are users who value time savings so highly that the reduction in congestion more than compensates for the imposition of prices. Among the losers are those people diverted to other less convenient facilities, modes, or hours and users of the alternative facilities or modes that are now, because of diversion, more congested.[11]

An important caveat is that the revenue generated by raising prices is retained for other purposes and not returned to the users. Economists have long demonstrated that gains offset the losses when this revenue is considered; in theory, the revenue could be used to compensate the losers for their losses. It may be difficult, however, to design a compensation scheme that is as concrete and credible as the losses the losing groups experience or anticipate. When economists speak of winners compensating losers, moreover, they rarely acknowledge that individual compensation is both impractical and undesirable. If one could compensate each individual for losses when raising prices on an existing facility, the prices would no longer affect travel behavior. Economists usually refer to compensating broad classes or groups rather than individuals. Thus, travelers as a group might be compensated for the burdens of pricing, say, through a general tax reduction, but many individual users are still likely to lose despite compensation.

Where prices are imposed on a new facility to recover its investment cost, by contrast, the compensation to users is visible and automatic. All users deem themselves winners—there are no losers. The users of competitive existing facilities and modes should benefit because congestion on these facilities should decline with the provision of new capacity. All users

11. For a more detailed discussion of who wins and loses under pricing see José A. Gómez-Ibáñez, "The Political Economy of Highway Tolls and Congestion Pricing," *Traffic Quarterly*, vol. 46 (July 1992), pp. 343–60.

of the new facility benefit as well since they find using the new facility, even with its prices, preferable to their old options. The users might consider themselves worse off only if they believe other options besides user charges, such as general tax revenues, are realistic possibilities for financing the new facility. This possibility can often lurk in the background, of course, as it did in several of the California highway privatization initiatives.

Even when privatization means the improvement of an existing facility or service, users and economists may disagree about the appropriate prices and profits. This disparity occurs because the public enterprises to be privatized typically understate their true costs in their financial accounts. Past investments financed by central government grants often do not appear on the public enterprise's accounts at all, even though those grants represent a real cost to society. Government equity investments often receive little or no recognition or return, as in the public sector's plan to extend the Dulles Toll Road in Virginia. Even investments financed by the enterprise's own debt or bonds may appear on the books only as long as the bonds are outstanding. Where such capital costs are carried, moreover, they are usually valued at historic costs rather than current replacement or opportunity costs, which are often higher because of inflation. There are exceptions: the British Airports Authority valued its investments at replacement cost even when it was publicly owned. For the Albany and Los Angeles airports and most other public airports, bus, or turnpike authorities in the United States and elsewhere, however, undervaluation is the norm.

Privatization is likely to lead to a revaluation of assets when capacity is expanded. At the least, the new private owners will insist on recovering the full costs of any capacity expansion they are required to make. Even if the old investments continue to be valued at historic costs, the inclusion of new investments in the rate base will force price increases. From the economist's perspective, moreover, prices that reflect replacement costs should be applied to users of old as well as new capacity, since a reduction in the use of the old capacity might reduce the need for new. This view often appeals to governments as well since, as in the cases of the Albany and Los Angeles airports, it may allow them to capture the return or inflationary windfalls on their historic investments to be put to use for other purposes.

Users or the general public, however, are likely to perceive revaluing old investments as unfair. Revaluation often would mean charging users more than the old public enterprise actually paid. To the economist, no excess profit or surplus occurs because the true economic cost is the replacement cost when the enterprise is expanding. In the popular mind, however, charging more than was paid is the definition of excess profit.

Popular and economic pricing prescriptions may not differ greatly if the facility generates important external benefits. If an expanded facility or service reduced congestion on competing facilities, for example, an economist might recommend a pricing discount similar to that implied by undervaluing the firm's old assets. Such lucky coincidences are more likely with buses or highways than with airports, however, since the latter usually generate external costs (for example, noise or air pollution) rather than external benefits (for example, congestion reductions on competing facilities).

It is no surprise, therefore, that most often wholly new transportation infrastructure, rather than existing facilities, is privatized. Virtually all of the highway privatization programs in the United States, Europe, and the developing world have been of new facilities. Airport privatizations, by contrast, have generally been slower, in part because existing public facilities are usually affected and thus price increases and profits are potentially more controversial. The British Airports Authority stands as something of an exception, perhaps because the former public enterprise was one of the few that valued its investments at replacement costs, so the expected price shock was smaller.

In short, surpluses or excess profits are usually as much of a political liability as an asset, especially if they are visible, large, and generated by price increases or wage cuts rather than pure technical efficiency gains. The general public and users are likely to see surpluses where economists do not, moreover, especially when existing infrastructure or facilities are privatized.

The Politics of Losses

When privatization involves losses, someone will have to make up the shortfall. The fundamental issue is who. There are two basic options: other private enterprises or the public sector. Often the two sources are used in combination, especially when the prospects for profitability are low, but each has advantages and disadvantages.

Other Private Enterprises

Other private enterprises are often the sources first considered, since one of the attractions of privatization is that it will not lay claim to scarce public resources. Support can come from several private activities. Private toll road projects around the world almost always look to construction companies with an interest in building the project to provide equity and other financing on more favorable terms than might be available in the normal capital markets. Financial institutions with an interest in underwriting the debt may be asked to contribute to project development costs. Some of the private toll road projects in the United States and elsewhere, and all the U.S. private rail projects, attempted to draw support from real estate developers and landowners whose properties might be enhanced. (A real estate developer was the originator of the private rail proposal in Boston.) And in two of the U.S. rail cases, rail equipment manufacturers were expected to assume a share of development costs and risks.

These efforts can be viewed as attempts to capture external benefits of the project. Real estate developers and landowners benefit, for example, because the toll road or rail line increases the accessibility and marketability of their developments. If the toll road or rail line operator owned all of the property that would be served, these benefits would automatically be internalized. But this is rarely the case, though the rail lines at Disney amusement parks illustrate the possibilities.

Most of the time, however, these external benefits are not the type that an economist would consider as warranting public subsidy since they are, in principle, usually capturable by the private transportation provider. Property values appreciate around a new rail line or road, for example, because travel times and costs are reduced for users. Thus, even if a toll road or rail line operator does not own all the property served, it could, at least in theory, capture the benefits of improved accessibility through the tolls or fares charged users. In practice, it is impossible for the transportation firm to vary charges enough by location or type of user to capture all of the benefits the users receive. Transportation firms are no different than many other private firms in this respect, however. It is impossible for a computer manufacturer, for example, to so discriminate in its pricing as to capture all of the benefits the different users of its product gain.

In essence, then, the private transportation firm's principal leverage for winning support from other private interests is usually an offer to share benefits that in theory it might capture for itself, but in practice the other parties might be in a better position to realize. To the extent that the other parties can capture benefits that the transportation provider cannot, both parties gain. Property developers, since they may be able to vary their prices by location, often capture more of the fine-grained accessibility advantages that a new road or rail line may offer. Then too the construction, financial, and equipment companies involved may be better able than private transportation firms to assess and control certain risks, or they may hope to capture other lucrative business because of their experience in a pioneering privatization project.

Private parties must believe that they will not capture the potential benefits unless they provide support for the project, which is often a problem when many small owners are involved. In DartRAIL, for example, property owners, believing they might realize benefits at no cost, had a strong incentive to hold back and not contribute in the hope that others would. Similarly, a construction company or equipment manufacturer with superior skill or products may hope to win the contract even if it does not support the project.

To demonstrate benefits and attract additional private support, augmenting the original plan in potentially costly ways is often necessary. Construction or equipment companies may want guaranteed contracts with negotiated prices rather than competitive bidding. To entice property developers' support, DartRAIL added a monorail loop servicing Tysons Corner; Boston's Seaport monorail added a station at Fan Pier; and MTI's first proposal included a possible spur line serving International Drive. Similarly, the proposed private Virginia turnpike extension may have been a bit longer than the public alternative to better accommodate adjacent private property owners whose help the private highway developers needed; the private developers of the proposed San Diego border connector road also made changes to garner private sector support.

Modifying the project to clarify the assignment of benefits can create something of a paradox. On the one hand, it is desirable to clearly assign benefits in order to quickly gain support. On the other hand, such clear assignment makes it easier for others, such as rival property developers, to see how the offering could be disadvantageous for them and therefore worth opposing. The pragmatic resolution of this paradox, as suggested in many of the cases studied, is to spread benefits more widely, diffusing opposition and broadening support, while simultaneously trying to redistribute system costs over this broader base. This tactic, however, raises new complexities, those of coalition building.

In building coalitions to gain approval for its projects, the private sector relies primarily on financial incentives—for example, by modifying system designs to confer more direct benefits or sharing revenues. In contrast, public agencies have a larger set of available tools than financial incentives. For example, they can accelerate project approval for those cooperating, offer density bonuses, and establish assessment districts. The public sector also has powerful means for securing participation even when not volunteered (for example, use of eminent domain to assemble sites, forcing cross subsidies, or the imposition of special benefit assessment districts).

Although limited in the tools at their disposal, the private parties usually do have at least one potential bargaining advantage: their negotiations are not as subject to public scrutiny. Consequently, they may have greater latitude in tailoring individual agreements and negotiating tradeoffs, for example, exercising greater discretion in station or ramp location and design or making above-fair-market offers for critical land assemblies. Moreover, private sector negotiators may engender more trust with private counterparts in the belief that they share a common profit motive and are not constrained by a mandate to secure public benefit. This sense of shared values may be important in assembling right-of-way.

As the circle of interests widens, however, the public sector almost inevitably becomes more participatory—even if the original initiative came almost exclusively from the private sector. Even without complications, any private infrastructure development usually must submit itself for environmental review and other site approvals. Public sector involvement, however, introduces new actors whose criteria for evaluating projects can differ markedly from those of the private sector.

Public Sector Support

Efficiency considerations might justify public sector support, most notably for achieving relief on competing congested facilities. Appeals for public subsidies are rarely couched in such terms, however. At least initially, the proponents of most private infrastructure developments typically characterize their projects as wholly private enterprises, whose costs will be absorbed entirely by the private sector.

This assertion usually can be questioned on several grounds. For example, many projects exclude the cost of publicly owned right-of-way they will use. They either assume that it will be provided in exchange for the service offered, or they simply choose to exclude an estimate from their budgets. Other public expenses (for example, planning, safety enforcement) required to support these projects are often ignored. If these costs are added to projects, financial viability weakens, raising concerns that the public could eventually be left to operate or dismantle them.

Even if no direct aid is requested, moreover, the public sector usually is mindful of a broader range of project impacts or concerns. The private sector's attempt to capture external value to support its offering may be perceived, for example, as something that simultaneously imposes costs on others. In the eyes of the public sector "the public" does not consist solely of system patrons or the owners of real estate developments. Rather, it extends to the taxpayers who might be required to eventually subsidize an offering, and to all area residents who might be affected by any increased traffic, pollution, noise, and density. The public, moreover, is composed not only of the present but of future generations, who will have to live with the consequences of a development.

To the extent that the proposed private system is something that the public wants, and is likely to build anyway, the private sector offer may be enticing. However, in at least one of the cases studied (the Orlando maglev), the private sector's motivation was apparently as much to demonstrate the feasibility of a new technology as it was to offer a public service; indeed, there was little indication of any local demand for the proposed system. Similarly, though to a somewhat lesser extent, Boston's World Trade Center monorail seemed to be as much motivated by the interests of certain real estate developers as by any broader public interest in the service. In some airport privatizations, such as Albany, private backers are driven at least as much by property development as by the potential for air service functions. Needless to say, to the extent that private offerings are unsolicited and not well aligned with any defined public need, they are likely to provoke public suspicion.

Requests for public support usually become explicit at some point. Often direct cash subsidies are not solicited but rather borrowings of public powers. Eminent domain may be needed to aid in corridor assembly, special tax assessment districts to compel contributions from the private property owners who benefit, or zoning variances to allow property development expected to cross subsidize the transportation project. These borrowings of public powers broaden the scope of potential private opposition and public concerns about the project.

To prove that public support is justified, the developer may have to enhance the proposal, increasing its public benefits. In Boston, the local transit authority wanted the World Trade Center monorail extended to improve connections between several existing transit lines. Similarly, Florida's Orange County wanted the maglev developer to participate in developing a transit system servicing International Drive, thereby supplying to the corridor what a public effort was to have provided. The

Virginia and California private toll road proposals have explored design changes to better accommodate environmental concerns; some have also proposed enhancements to better integrate their facilities with existing public highway systems. Like the enhancements to attract private support, these modifications often increase system costs and further undermine profitability.

The public sector, moreover, is not a monolith. Agencies charged with providing transportation may hold viewpoints different than held by those responsible for environmental protection or economic development. National and local agencies may differ in their perspectives and, indeed, may be at odds over how to proceed in a specific case. In Boston's monorail project, the local redevelopment authority was primarily interested in developing the harbor area as an adjunct to and extension of downtown Boston, while the transit authority viewed harbor development as an opportunity to rationalize and improve public transportation. In Florida, similarly, at least some of the early delay of maglev was attributable to a need for the state and local governments to sort out their responsibilities for high-speed rail development and experimentation. The California Midstate highway may never be built because of the sharp differences in attitudes among neighboring communities on growth and environmental issues. Airport privatization proposals often reveal substantial differences among the different agencies responsible for air operations, for promoting economic growth, and for environmental matters.

Each sector has its own time horizons. The private sector wants to predict or control timing of investments, often with respect to cycles of financial and real estate markets. Some government officials may be anxious to demonstrate the merits of a certain program, such as privatization, and stress the importance of moving approvals along quickly. More typically, however, the public may be suspicious that haste leads to inadequate scrutiny of project impacts. The impact of these differing time horizons was illustrated by the attempt of the World Trade Center monorail proponents in Boston to portray their offering as an "interim transit solution" when it became clear that the project was facing stiff public opposition. Federal transit officials found the interim label more than reasonable, in part because they wanted an immediate demonstration of privatization. The reaction of state officials, however, was that few transport undertakings are ever interim; infrastructure tends to be durable and thus presents problems when system integration eventually occurs.

Dynamics

The relationship between private developers and their counterparts in the private and public sectors is also obviously not static. The dynamic forces at work alter and sometimes undo offerings. On the one hand, the private sector provider may need to link the offer with other development activities to cover costs. At the same time, the public sector is constantly trying to insure that the proposed project does not end up creating more public costs than benefits. Attempts to balance these diverse goals can create an ever-expanding circle of partners and parties at interest, including adjacent landowners, other private sector investors, and potential competitors, along with state and local agencies and neighborhood activists. Creating and maintaining an effective coalition, while keeping detractors at bay, can thus be immensely difficult. In several cases, the strategies utilized by project proponents sometimes failed to fully appreciate the complexity of the process, so that they periodically had to regroup and adapt alternative strategies.

Frequently the private developer initially tries to deal with these dynamics by attempting to limit or streamline the public review process. While an element of naivete or excessive optimism, especially about costs, may affect the initial choice of strategy, a desire to limit a project's exposure to public agency review and involvement also influences the developer. The project teams typically include people familiar with the public review process and the lengthy delays and additional proposal development costs that often result. By limiting public participation, proponents hope to maintain maximum control over project timing and costs. Often project proponents may seek initial project approval from what they perceive to be the highest necessary level of public authority, for example, the rationale being that such high-level approval accelerates processing at lower levels. However, this move may alienate other public sector actors and make later relationships more complex. Moreover, the private backers almost invariably find themselves drawn into increasing public sector review, first by virtue of securing permits and right-of-way, and then because of a desire to use public sector tools to create their coalitions and to finance their offerings.

The failure of initial strategies typically forces project developers into ever more complicated, time-consuming, and costly negotiations aimed at structuring partnerships and building public and private sector support. Ultimately, the costs of building and maintaining coalitions with significant private sector participation may exceed expected efficiency gains, especially savings projected from more rapid completion of the project. Greater complexity increases costs, which may cause the private sector to abandon its efforts. Alternatively, the projects may end up evolving into quasi–public-private partnerships, both in the land development and the transport function, rather than straightforward privatization efforts. In either case, the eventual outcome is likely to be rather different from that initially envisioned by proponents. Furthermore, complete reversion to a public sector activity cannot be ruled out, as the French and Spanish highway experiences amply illustrate.

The tendency for the more complex privatization proposals to evolve into public-private partnerships or joint ventures raises the question of whether it might have been more sensible to recognize their ultimate quasi-partnership character at the outset. This thought underlies the suggestion often made, and in some jurisdictions accepted, that for highway projects government should assemble all the right-of-way and environmental permits before contracting with a private developer to construct and operate a facility. While this procedure may be more efficient, it may also stifle one of the more attractive features of privatization, that of private sector innovation and new ideas.

None of these problems is necessarily disabling or fatal. They do indicate, though, that some types of infrastructure privatization are likely to be more acceptable or doable than others. Perhaps somewhat surprisingly, both too little and too much profitability can be disabling to a privatization. Excess profits unleash the populist politics of equity and distribution, often inducing a stalemate between conflicting interests. Insufficient returns not only make strictly private financing difficult but also can induce a degenerative and self-destructive political dynamic. The economists' happy mean of "normal" returns, just sufficient to keep capital committed, thus may be beneficial politically as well as economically.

Part Four
Conclusion

Chapter 15. Lessons from Transport

AN OBVIOUS QUESTION IS WHETHER the generalizations that emerge from these analyses of transport privatization have any wider application, extending to other sectors of transport and other industries. In this chaper we explore that question.

Defining Success and Failure

We define success and failure broadly in this study. One obvious measure is the extent to which the basic economic or financial goals that typically motivate privatization can, at least in theory, be realized. Equally important, however, is the degree to which privatization is politically acceptable and therefore doable.

The economic or financial motives for privatization vary somewhat according to the type of privatization. The belief that the private sector is intrinsically more efficient than the public sector often motivates both the sale of state-owned enterprises and contracting out. A private management, motivated by the possibility of profit, it is hoped, will have stronger incentives to control costs and thereby reduce or eliminate the need to support the former state-owned enterprise or state-supplied service with scarce government tax revenues.

Efficiency can be defined more broadly, however, to include more efficient prices or innovative products and services. A private enterprise may have stronger incentives and greater latitude to tailor the prices of different services to more accurately reflect costs. This practice is generally regarded as desirable since it encourages customers to consume more wisely. Being more directly dependent on customer revenues, a private enterprise may also be more innovative and customer oriented in designing the services or products it offers. These pricing and product improvements can help further reduce taxpayer burdens, by making the private firm more profitable and also taxable, as well as improve the overall social performance of the industry.

In the case of private development of new infrastructure, the primary motive is typically the desire to tap new sources of funds to supplement the constrained resources of the public sector. Efficiency may still be an important and hoped-for advantage; the private sector may build infrastructure cheaper or faster, for example, or develop innovative designs or pricing. But the primary concern is usually that the public sector simply does not have the financial resources to build all the infrastructure needed. There are usually competing claims on tax receipts, and the government may perceive that raising taxes or increasing public indebtedness is politically unacceptable.

A third motivation for privatization, which applies primarily to the sale of existing state enterprises or infrastructure facilities, is the revenue the government hopes to receive. Of course governments can realize significant proceeds from such sales only if the enterprise or facility has the potential to generate operating income in excess of its operating expenses and investment needs. Even then, the cash flow may not be sufficient to recover the government's original investment in the facility. However, any recovery whatsoever is often attractive, especially when government resources are scarce and social needs pressing.

While the economic goals may appear attractive to governments in the abstract, their achievement in practice often creates economic losers as well as winners, and thus opposition and controversy. Privatization also can intensify environmental or other community concerns. Successful privatization therefore requires seeking out those limited opportunities where economic gains can be made at an acceptable political cost.

Indeed, the experience with transport suggests five conditions favorable to successful privatization: effective competition, large efficiency gains, few transfers, limited environmental problems or other externalities, and reasonable but not excessive profitability. None of these conditions is essential, and some are more important in certain economic or political milieus than others. Most often, however, these conditions are not only helpful but usually critical.

The Role of Competition

A competitive market environment is cited as important or even essential to successful privatization in almost every serious study of the subject yet done, irrespective of sector.[1] Where effective competition can be established and maintained in the relevant markets and activities, privatization has great potential to reduce costs and improve the quality of product or services; without competition, privatization and related reforms may bring some improvements, especially one-time gains in efficiency, but the gains are usually limited in scope, distribution, durability, and vitality.

Competition also helps insure that the private sector passes savings on to users and reduces popular suspicions about potential monopoly abuse. In general, a private firm might be presumed to be more tempted than a public agency to exploit any market power that it might possess. As long as a market is competitive, private firms cannot price much above their long-run marginal costs; they may be able to do so in the short run if demand temporarily outstrips supply, but only for as long as it takes to build additional capacity. If the market is not competitive, however, a firm may be able to sustain prices in excess of marginal costs, if politically permitted to do so. The history of government regulation attests to the difficulties of denying that permission. Without competition, users and public officials are likely to oppose privatization unless it is accompanied by government regulation of prices and service quality. Unfortunately, even when effective, regulation seldom completely allays public fears that a monopolist's prices are above costs or service quality is too low.

Privatization, therefore, tends to be most effective when the market is big enough to sustain more than one or two efficient suppliers, with relatively easy entry. Failing that, high transportation costs or other barriers to entry or supply from external sources should be minimal. Privatization thus makes more public policy sense when accompanied by free trade than by protectionism, and by aggressive antitrust policies, especially in smaller countries or regions with smaller internal markets.

Scale of market, moreover, does not necessarily increase with economic development. Some markets expand with development while others contract. For example, as observed in part I, the market for public transit is usually smaller in developed than in less developed countries, and the prospects for effective competition are thus usually greater in the developing world. Indeed, a striking feature of bus transit in the major cities of developing countries is the large number of small private opera-

1. For example, see John S. Vickers and George K. Yarrow, *Privatization: An Economic Analysis* (MIT Press, 1988); Raymond Vernon, *The Promise of Privatization: A Challenge for U.S. Policy* (New York: Council on Foreign Relations, 1988); and John Donahue, *The Privatization Decision: Public Ends, Private Means* (Basic Books, 1989).

tors, at least where local governments have not severely restricted entry. Bus ridership is usually much higher per capita in the cities of the developing world, and these high passenger volumes help make a large number of competing firms financially viable, while the high bus frequencies on many routes may reduce the advantages of the regular schedule or coordinated route network that large firms can offer.

Competition may conflict with other goals, however, especially financing new infrastructure from user revenue rather than scarce tax receipts or generating profits that the government can use for other public purposes. The British government is alleged to have sold its three London airports as a group rather than separately to maximize the prospects for a successful sale. And some developing countries have granted private bus companies exclusive franchises so that they can cross subsidize "socially desirable" but unprofitable routes with earnings on protected monopoly services.

Competition is also often difficult to secure in the case of infrastructure developments, where economies of scale, immobility, and siting problems tend to create a "natural" monopoly. Many countries that allow private toll roads thus require that a free alternative be available. But too competitive an alternative can threaten the profitability of the toll road; alternatively, if toll rates are too high, they can cause a serious misallocation of traffic between the tolled and untolled alternatives.[2]

Regulation can be used to prevent excess profits in natural monopolies. Regulation is not only a source of potential political controversy, however, but poses its own threats to efficiency. On the one hand, public officials may yield to popular pressures to set unrealistically low prices; these may lead, in turn, to service inadequacies and shortages. The most obvious examples are several third world cities where the bus industry has recently been privatized but not deregulated for fear, probably misguided, of monopoly abuse. Dissatisfaction with inadequate service, usually blamed on the companies rather than on fare regulators, may eventually lead to a renationalization of bus service and a fruitless repetition of the cycle of private and public ownership. Still another example is provided by the refusal of the French Ministry of Finance to agree to a formula for future toll increases to compensate for inflationary cost increases; this stance discouraged private investment in new private toll roads in France for at least a decade.

On the other hand, public officials may be captured by the firms they regulate and set prices or allowable rates of return well above costs. The

2. See chaps. 7 and 14.

Spanish formula for toll increases may be overly generous to some private toll road companies, as may be the rate-of-return caps established in the California toll road concession agreements.

Striking a balance between protecting the public and providing investors an opportunity to earn an adequate rate of return is not easy, even where regulators understand the need to do so.[3] Among the most attractive options is the *RPI* minus X regulation pioneered by the British in the 1980s and applied to its privatized airports and California's one-time contractual approach of setting a target rate of return in its toll road franchise agreements. Even these comparatively simple regulatory schemes require some sophistication to implement well, however, and thus the usual choice in the case of natural monopolies is among imperfect markets, imperfect regulation, or imperfect public enterprises.

Efficiency Gains versus Transfers

Privatization is almost certainly a more attractive public policy where the potential efficiency gains are great. Simply put, the larger the efficiency gains from privatization, the greater the prospects that most parties will gain and few will lose. The prospect of greater efficiency gains may also mean less pressure to extract economic rents from others, extractions which lead to regulatory or distributional problems that are better avoided. In short, the larger the efficiency gains, the larger the likely proprivatization constituency.

If there are significant transfers or redistributions of resources among parties, by contrast, one's position on privatization may depend in part on whether one is more sympathetic to the winners than the losers. The larger the transfers, moreover, the more controversial privatization is likely to be.

The Distinction

Not too surprisingly, efficiency gains and transfers are often confused with one another in privatization debates. The cost advantages claimed for privatization are sometimes transfers from one group to another

3. See Harvey Averch and Leland L. Johnson, "Behavior of the Firm under Regulatory Constraint," *American Economic Review*, vol. 52 (December 1962), pp. 1052–69, and the reviews of the literature in Alfred E. Kahn, *The Economics of Regulation: Principles and Institutions*, vol. 2, *Institutional Issues* (MIT Press, 1988), pp. 49–59; and William J. Baumol and Alvin K. Klevovick, "Input Choices and Rate of Return Regulation: An Overview of the Discussion," *Bell Journal of Economics and Management Science*, vol. 1 (Autumn 1970), pp. 162–90.

rather than real resource savings for the economy as a whole. For example, private bus companies often pay lower wages than the public bus companies they replace, especially in Britain and the United States.[4] Lower wage rates reduce budgetary costs but, without productivity increases, do not reduce the labor resources required. To the extent that the lower input prices paid by private vendors are closer to true free market prices, as presumably they often would be, then a more efficient combination of inputs should be achieved by the private than the public sector. In short, working with better input price signals, the private supplier should be more productive than a public sector counterpart, all else equal.

Similarly, landowners and developers might be more willing to donate right-of-way to private than public road projects. Most of the land for the private Dulles Toll Road Extension, for example, would be donated by neighboring landowners who stand to benefit. Landowners also commonly donate land to public projects where donations might encourage earlier construction of an adjoining road, although the threat that a project might not survive without donations may be more credible where a private rather than a public operator is involved. Whether public or private, however, donations of land represent a transfer from landowners to users or investors and generally do not reduce the total land required for the project.[5] Furthermore, if the land input is fixed, then no productivity gain would be expected even if the private sector faced more realistic market prices for land.

Another transfer often misclassified as a cost saving is the differential tax treatment of private and public enterprises. In this case, the cost differential usually works to the disadvantage of private enterprises, which generally have higher tax liabilities. In the United States, public toll road operators generally can issue tax-exempt debt and do not pay corporate income or property taxes, unlike their private counterparts. If the nominal cost of public debt in the United States is lower than private debt under current tax laws and market conditions, then the use of public rather than private debt simply transfers some burden to the federal, state, and local taxpayer. In essence, tax receipts are reduced to the extent

4. In the developing world, the wage differential between private and public bus operators is usually less pronounced. In some privatizations in the United States and Europe, moreover, wage differentials are not allowed; for example, California has specified that private toll road builders must pay the same prevailing union wages as public authorities.

5. Interestingly, the private version of the Dulles Toll Road Extension apparently would require slightly *more* land and be longer than a public alternative in order to better accommodate adjacent landowners and thereby garner their cooperation and contributions.

that the total supply of tax shields is increased in the economy. The tax loss commonly will be captured by investors, in higher returns, or by facility users, in lower fees or tariffs. If nominal public financing costs are also lower because a state or local public agency contributes hidden equity (for example, initial planning costs or working capital) that earns no return, then the state or local taxpayer also loses (from not receiving a fair return on that equity) while the facility user usually benefits (in lower tolls).

Evidence of Efficiency Gains

The transport case studies provide strong evidence of private sector cost reductions in labor-intensive services. The studies of bus privatization or contracting out in Britain and the United States show fairly consistent savings of about 20 percent, even when public sector costs for monitoring the private firms for contract performance are considered. Savings can be even greater for privatizing buses in developing countries. Less than half of these savings usually originates from lower wage rates; the bulk comes from higher labor productivity, leaner management and overhead costs, and reduced maintenance expenses. These results are similar to those found in numerous previous studies of labor-intensive services in other sectors, such as garbage collection or building maintenance, at least as long as there is competition to insure that the private operators remain efficient.[6]

The transport experiences also strongly support the proposition that private firms are likely to be more innovative and market oriented in their services and facility designs. This result holds, moreover, for both labor-

6. In a comparison of public and private provision of eight different labor-intensive services, Barbara Stevens found that the cost savings stemmed largely from higher labor productivity and not just from lower wage rates in seven of the eight cases. See Barbara J. Stevens, "Comparing Public and Private Sector Production Efficiency: An Analysis of Eight Activities," *National Productivity Review*, vol. 3 (Autumn 1984), pp. 395–406. John Donahue, in reviewing comparative cost studies, concludes that the critical factor is not the form of ownership but the presence of competitive markets. He writes that public versus private market matters in most cases, but competitive versus non-competitive market usually matters more. One exception to the pattern of finding lower costs with privatization is the literature on the comparative costs of publicly and privately owned electric, gas, water, and sewage utilities. There is no consensus in these studies on which form of ownership has lower costs. The privately owned utilities are usually publicly regulated, however, and public regulation can reduce the potential efficiency advantages of private ownership. See John D. Donahue, *The Privatization Decision: Public Ends and Private Means* (Basic Books, 1989).

intensive services and infrastructure. In bus transport, competition has stimulated major service innovations, such as greater use of minibuses and express services, in developed and developing countries. In infrastructure, private toll roads are generally more innovative than their public counterparts, especially in design and toll collection. Indeed, as already suggested, enhanced innovation is one of the most dependable benefits of privatization, possibly justifying some privatization even where short-run static costs are thereby enhanced.

The transport case studies provide only limited evidence, however, that the private sector can build, as opposed to operate, infrastructure more cheaply or quickly than the public sector. The only careful comparison of private and public infrastructure costs in our transport case studies, from one of the French toll road companies, suggests that private construction may be 20 percent less expensive. This one case aside, however, there is little direct evidence that private providers of infrastructure are less costly than public. As a result, any comparison of private and public infrastructure costs must rest mostly on a subjective assessment of the claims of various supporters and detractors, with only limited empirical evidence for guidance.

Proponents of privatization argue that private firms, motivated by profit, should be able to build infrastructure more cheaply or quickly by avoiding cumbersome public sector bidding and contracting requirements. The public sector generally plans, designs, bids, and builds major facilities in a sequential process, completing each stage before the next is begun. Private firms may have more flexibility to use design-build or fast-track parallel processes in which design engineers and private contractors are selected simultaneously and the planning, designing, bidding, and construction phases overlap. Faster construction saves on the capital required for a project by bringing the investment into service more quickly.[7]

The private sector may be subject to other sources of delay, however, that at least partially offset the potential advantages from more flexible

7. The public sector could use similar procedures. Many states use overlapping design-build procedures to speed the construction of prisons. But experience suggests that the public sector is less likely to be motivated by these time and cost savings than the private sector; conversely, the public sector may be more sensitive than the private sector to the mistakes that may occur because of haste (for example, fraud or overlooking environmental concerns). In general, an asymmetry may exist wherein the public sector is more exposed by hasty decisions and processes than the private sector.

procurement. Negotiation of the regulatory scheme or concession agreements may delay private projects. Similarly, private firms may have more problems in securing environmental and other permits than their public counterparts. Furthermore, arranging financing can be time consuming, especially when the project is the first of its type or potentially controversial subsidies are requested. All these problems have been evident in U.S. private highway and rail projects; backers of the Dulles Toll Road Extension originally planned to begin construction in two years, for example, but after six, construction still had not begun.

The private sector, though, may be better able to exploit economies of scale, scope, and experience in construction and operations. In building or operating a number of facilities in several locations, private firms may achieve greater specialization of labor by being able to afford experts with specialized technical or managerial skills, while smaller public agencies have to rely on generalists or outside consultants not fully familiar with the proposed installation. Multiple-plant operation may also allow the private operator to achieve economies in administrative or overhead functions and to offer staff more opportunities and incentives for career advancement, thereby enabling the recruitment of a better work force at less cost, all else equal. Private operators may also be better positioned to exploit their experience, or the learning curve, since by building larger facilities or building more often, they do not have to learn about the practical and technological problems afresh each time. Many of the transport cases offer strong evidence of these economies, such as the emergence of bus holding companies in Britain or the involvement of large European toll road operators in backing private toll road projects in the United States. The public sector may encounter difficulties in achieving similar economies of scale, scope, and experience internally. Even by banding together on a regional basis, local communities are unlikely to build or operate more than one large facility every decade or two.

The public sector's recognition of these potential economies is reflected in the popularity of contracting with private firms to construct complex infrastructure facilities, even when they are publicly owned, and in the growing practice of contracting for operational management. Sometimes the private sector's participation in a presumably public activity may already be so extensive that the cost savings from further privatization are limited. This seems most plausible for infrastructure construction, given the nearly universal practice of using competitive bids to contract with large and experienced private companies. The principal gains from privatization of infrastructure may therefore lie in more efficient operations and innovative designs and services than in cheaper construction costs.

Winners and Losers

Privatization almost always results in some net redistributions or transfers, although in some cases all parties seem to win. The actual incidence of gains and losses from privatization depends considerably on the particulars of the case. Nevertheless, some broad tendencies can be identified, even if only tentatively.

The most likely and common losers from privatization are organized labor. Labor will lose to the extent that private sector operation results in lower wage rates or less protective work rules than those found in the public sector. Labor does not always lose, however. Wage rates may be protected as one of the conditions of privatization, as with California's requirement that private roads pay prevailing union rates, or through a two-tier wage system, as in Britain's privatized bus industry. Privatization may also lead to increases in total employment in the industry if, for example, more infrastructure can be built by supplementing scarce tax resources with private financing or if the service innovations and efficiency improvements are so great as to significantly increase service demand.

Another possible loser is the local landowner. A private operator may be more successful than the public sector in extracting land donations or other contributions to advance the enterprise. Even so, landowners still may not lose, particularly where the infrastructure would not be built as soon or at all without private financing, and indeed, in these situations local landowners are often strong backers of private infrastructure projects.

The clearest winners from privatization are normally taxpayers. Privatization is almost always undertaken because tax resources are thought to be limited, and thus privatization is nearly always designed to cut the public subsidies needed by former state-owned enterprises or contracted services, to realize some proceeds from the sale of public assets, or to relieve the pressure to finance new infrastructure with tax receipts. In nearly all the transport cases studied, these benefits have occurred to some degree, although sometimes not as much as initially hoped.

Taxpayers can also gain because private activities are usually subject to more tax liabilities than the public activities they replace. Similarly, to the extent that private equity or debt replaces public equity that receives little or no return, taxpayers gain from not having to make that uncompensated contribution. At the same time they also gain from any higher income tax paid on returns realized on the private equity.

Private investors might gain from privatization if they are able to retain any economic rents, captured from labor or landowners, or efficiency gains from privatization, instead of passing these on to facility users in the form of lower charges or better services. The prospects for doing so depend on how competitive the markets for the facility's services might be or, failing that, how well public regulators do their jobs. In a competitive market or under perfect regulation (a perhaps unattainable ideal), the facility owners would be forced to pass these savings on to facility users and would earn only a normal return on their investment. With a less competitive market, or compliant regulation, the facility investors might be able to earn above-market returns.[8]

Whether users of a new service or infrastructure facility would gain or lose from privatization depends on the situation. They are most likely to lose in three circumstances: first, where the transfers to taxpayers are large, because cutting subsidies or realizing sale proceeds is an important goal or because a private firm faces much higher tax liabilities; second, where the efficiency gains or the rents captured from landowners or labor are modest; and third, where lack of effective competition or regulation allows the private firm to retain any cost savings or transfers rather than pass them on to users in lower prices or expanded service.

When users lose in transport privatization, they usually do so through a combination of the first two circumstances (that is, where the transfers to taxpayers, intentional or not, are larger than the efficiency savings and the transfers from labor or landowners). This happened to some degree with British buses, for example, especially in several metropolitan areas where the public bus companies had been heavily subsidized before privatization. Privatization softened the blow of subsidy reductions by cutting unit costs, but not enough to make all British bus users whole. The Dulles Toll Road Extension is another example, although only if one believes that a public toll road was a realistic possibility. On the Dulles road, the users might have gained from increased landowner contribu-

8. With excessively stringent regulation the investors could lose by earning a below-market return in the short run but would eventually withdraw their capital by underinvesting or not maintaining the facility.

tions to the private project, but privatization seemed to offer few construction cost savings, and the private company would not enjoy the tax-exempt status or the uncompensated public equity of the public alternative. Some of the airport privatization proposals are also of this type, especially where the government's goal of realizing sale or lease proceeds is likely to offset any efficiency gains.

Lack of effective competition or regulation, by contrast, is a less common source of user losses for privatized services. Bus competition was almost invariably effective in the cases studied, the most notable exceptions being one of the British metropolitan areas, where a large and well-entrenched public firm was privatized intact, and Santiago, Chile, where the private minibus route associations functioned as an effective cartel once prices were deregulated. Indeed, users are most likely to lose with bus privatization when regulators are so intent on protecting the public from fare increases that they bankrupt the bus companies, thereby disrupting or even destroying service.

Users gained significantly from privatization, moreover, where privatization meant a new facility or service not otherwise or previously available, or that the public sector was unwilling or unable to finance through tax resources. Many of the new or proposed private toll roads in the developing world, Europe, and the United States fit this model. Similarly, users unequivocally gained in the cities of the developing world when private bus service expanded because a shortage of subsidies had led to inadequate public bus service. Again, a striking lesson of the transport privatization experiences is that product or service innovations often may be as important a source of efficiency gains for users as simple unit cost reductions.

The Disadvantages of Externalities

The presence of significant externalities is usually a disadvantage for privatization in two respects. First, it can lead to setting prices at levels other than those needed to recover the private enterprise's costs. Large positive externalities, such as congestion reductions on competing but underpriced facilities or modes, may justify pricing below the enterprise's costs; large negative externalities, such as noise or air pollution, can justify pricing above the enterprise's costs.

Second, local neighborhood and environmental opposition can be an important bar to the siting of new infrastructure facilities, whether public or private. Siting problems reduce the chances that a private facility, once built, will ever face competition, and thus encourage price regulation and its attendant problems. Moreover, siting controversies may be an important barrier to building the private facility in the first place.

While new facilities are sometimes welcomed as spurs to development, especially in rural or outlying areas, in developed countries these situations may now be the exception rather than the rule. In the United States, and, to a slightly lesser extent, in Europe, each new decade has added new sources of concern and new government regulations controlling infrastructure siting decisions.[9] By the 1960s, for example, U.S. neighborhoods in the path of new highways had learned to mobilize politically; their opposition eventually led to the cancellation of some proposed metropolitan expressways and to federal requirements that highway planners consider "no build" and mass transit alternatives. By the 1970s, concern about automobile air pollution and the destruction of parklands and other sensitive environmental areas led to mandatory environmental reviews and public hearings on highway and other major infrastructure projects. The 1980s also brought a concern that new highways would stimulate too much development, especially in suburban areas where growing traffic congestion and development densities seemed to threaten the quality of life (as suggested by some of the opposition to the California highway privatization initiatives).

It is unclear whether a private operator has any major advantages in resolving siting or environmental controversies. On the one hand, private firms may have more flexibility than public agencies in the compensation they can offer objectors, more incentive to compromise to avoid the financial costs of delay, and a greater ability to avoid the public spotlight and controversy while resolving concerns of objectors. On the other hand, private firms may be deemed less likely to take their environmental and community concerns seriously and their sensitivity to the financial costs of delay may lead them to avoid or abandon controversial projects.

Ultimately, public agencies may have an advantage over the private sector simply because they have more established institutions and proceedings for dealing with the types of equity issues that arise in settling conflicts over externalities. The private sector, almost by definition, has to rely on bargaining to reconcile conflicting interests. Where the parties are numerous or the conflict is so polarized that mutual agreement seems

9. For a description of the evolution of public concerns and government regulations governing highway siting decisions see Alan A. Altshuler, *The Urban Transportation System: Politics and Policy Innovation* (MIT Press, 1979).

difficult if not impossible, public institutions for conflict resolution, with their established procedures and authority, may be quicker or their participation unavoidable.

Whether the involvement of a private firm on balance reduces or increases siting or environmental problems therefore depends on the circumstances, such as the type of facility, the reputation of the private firm and its skills at negotiation and compromise, and the strength and nature of the local opposition. Privatization is clearly easier to implement where externality problems are relatively minor.

Avoiding Subsidies and Surpluses

Finally, privatization is usually easiest when the prices charged users just recover the costs, including a normal return on investment, of the private facility or service provider. The basic problem with surpluses or subsidies is that they are another form of societal redistribution or transfer and thus, as noted earlier, often generate controversy.

Subsidies can arise from many different considerations. The most common is simply that the privatized enterprise, no matter what its efficiency gains might be, cannot earn enough to attract needed capital, as is usually true of urban rail transit. If prices are regulated to minimize the impact on certain consumer groups, often a temptation with urban buses, then this possibility of insufficient return on capital becomes even more likely.

Financing the subsidy can be a serious problem, especially when privatization is taking place. Governments are usually interested in privatization because tax resources are scarce; they are unlikely to be interested in subsidizing a private firm unless the only politically feasible alternative is larger subsidies to a public firm, as is true of some urban bus services in developed countries. Other private entities might be induced to provide subsidies, especially landowners who stand to benefit from a nearby private road or rail project. The U.S. experience demonstrates, however, that landowners have strong incentives to hold out if they have any expectation that others might contribute instead, or they may demand in return modifications that disadvantage their competitors.

Surpluses can be advantageous for governments if they can be captured, in sale or lease payments, for example, and used for tax relief or other public purposes. Users are likely to object to such transfers, however, as happened with proposals to privatize airports. Surpluses retained by the private investors will usually raise even stronger objections, including charges that the government has failed to protect the public from monopoly.

In this context, schemes where some users cross subsidize other users are often attractive. Cross subsidy eliminates the potential drain on the public treasury and makes the transfers less visible and thus less controversial. The French have used cross subsidies within their autoroute companies to finance the construction of unprofitable toll roads, while a few cities in developing countries use such schemes to support unprofitable bus routes.

Cross subsidy has disadvantages, however, as illustrated by roads and buses. A troubling administrative problem is insuring that the less lucrative as well as the more profitable products or services are both provided, since there are obvious incentives to skimp on the less lucrative. The monitoring requirements for successful cross subsidies thus can be a major strain on government departments, especially for bus services in developing countries. Government officials also have less incentive to question whether the subsidies are warranted since there is no direct burden on the public purse. The expansion of the French autoroute system has been dictated largely by the profitability of the main trunk routes, for example, rather than by any careful calculation of the social benefits and costs of new branches.

This last example raises the issue of the extent to which subsidies, even if politically acceptable, undermine the incentives for economic efficiency usually provided by privatization. The experience with contracting out of subsidized bus services in Britain and the United States demonstrates that incentives for private firms to control costs can be maintained despite subsidies. In both countries, services were procured through competitive bidding and the resulting unit cost savings were much greater than the cost of monitoring contractor performance. In Britain, contracting for subsidized services promoted competition in commercial (unsubsidized) services as well, by providing a niche for small firms to enter the market.

More often, however, subsidy programs are not carefully enough designed or implemented to maintain efficiency incentives. The experience with contracting for dial-a-ride bus services in the United States has been disappointing, in part because public officials have often specified specialized equipment or other terms that limited the number of competing firms. Overly generous loan and exchange rate guarantees may have reduced the incentives of some of the early Spanish private toll road companies to monitor costs, and the possibility of renegotiating concession terms and other aid in the event of construction cost overruns may have had similar effects in Mexico.

In sum, the success of a privatization initiative, broadly defined to include political as well as economic feasibility, seems most likely when the activity being privatized sells into a competitive market, substantial efficiency gains can be garnered from privatization, the redistribution or income transfers created by the privatization are not overly large and complex, the activity to be privatized does not lead to extensive externalities or spillovers, and major subsidies are not required. The attractiveness and viability of privatization can also depend on the stage of economic and political development, as the earlier discussions of both urban buses and private toll roads illustrated. Privatization thrives best when these dimensions are near some golden mean: too much or too little economic development and profitability are likely to be troublesome.

Applying the Lessons More Broadly

The lessons learned in our case studies of transportation privatization can be extended to other sectors of transport and to other industries. To illustrate, consider two activities commonly proposed for privatization in the United States: solid waste disposal (for example, landfills and incinerators), and toll roads. Applying the principles already enunciated, waste disposal should be a good deal easier to privatize than toll roads. To start, waste disposal tends to be a more competitive activity than toll roads. Almost any community in the United States will have more than one waste disposal firm willing to bid on a local contract or franchise, and in many instances more than one operator can be licensed to serve the same geographic area without a disabling loss of efficiency, mainly because costs of transporting waste to the disposal site is an increasingly less important percentage of total disposal cost. In many smaller U.S. communities, government has never provided organized public waste disposal services; instead, private disposal firms have competed directly for the business. By contrast, a toll road is almost invariably linked with at least some aspects of local monopoly.

Substantial efficiency gains are likely to be realized by the privatization of waste disposal as well. The rise of the environmental movement in the late 1960s dramatically changed waste disposal practices and, in the process, transformed private waste disposal from a labor-intensive industry composed of many small firms into a capital-intensive industry, dominated by a handful of giants. The same economies of scale, scope, and learning that drove consolidation of the private industry have also steadily opened up a gap between the costs achievable by the increasingly large private firms and those achievable by small- to medium-sized public agencies doing trash disposal. Indeed, where communities remain wary of relying on private solid waste firms, they have often banded together to form special districts to build and operate their own public waste disposal facilities, thereby giving these communities some of the economies of scale exploited by the large private firms. Even so, a group of communities is unlikely to exploit the full economies of experience, since they are unlikely to build a major facility more than once or twice every generation. Not surprisingly, more than half the U.S. cities studied in a recent large survey rely on private waste disposal.[10] Although private landfills represent only 14 percent of the total number in the United States, they contain about half of the nation's existing disposal space.[11] This privatization trend has been driven largely by the substantial efficiencies achievable through privatization.

By contrast, the efficiency gains from privatizing the ownership or operation of roads are more modest. Because of economies of experience, scale, and learning, major roads in the United States have long been built by large private construction firms acting under contract from public agencies. Heavy road maintenance and often even light road maintenance have also long been handled by outsourcing to the private sector. Policing and related public safety activities might be expected to be retained in the public sector, whether a road was privately or publicly owned. Accordingly, the only important sources of efficiency gain in road privatization are in operating costs, which are a fairly modest proportion of total costs, or in the identification of an overlooked route or an innovative pricing strategy.

Any redistributions or income transfers brought about by privatization of trash disposal should also be modest. Redistributions from labor, if any, would be small since waste disposal, as opposed to waste collection, is more capital than labor intensive. Any objections from labor can usually be finessed fairly inexpensively if the privatized trash remover agrees to employ any displaced public employees. Highways, however, often effect substantial redistributions of wealth and income, especially by the choice of alignment.

10. American Public Works Research Foundation, *Private Sector Contracts and Agreements for Solid Waste Collection and Disposal Services* (Chicago, Ill.: American Public Works Research Foundation, 1990).

11. National Solid Wastes Management Association, *Privatizing Municipal Waste Services* (Washington: National Solid Wastes Management Association, 1989), p. 5.

Toll roads are also likely to face much more complicated environmental or externality problems than waste removal. Toll roads need long linear strips of property, thereby obviously affecting several neighborhoods. Waste disposal activities, by contrast, arouse only the neighborhood where a landfill is to be located. Indeed, waste disposal is a classic case of the so-called not-in-my-backyard syndrome. While those people near the site of a proposed landfill almost always ardently oppose it, those any distance away and served by that landfill usually will be in favor. The externality problem therefore is more focused or concentrated for waste disposal than for toll roads.

Finally, solid waste disposal is usually a fairly profitable activity because of the growing volumes of trash and a shortage of modern disposal facilities. These profits give the private waste company the wherewithal to compensate the communities in which they locate. By contrast, new private toll roads in the United States often must seek contributions from landowners or local communities along their alignment since the U.S. expressway system is already so extensive that most of the roads that might be profitable from tolls alone have already been built. Putting all these considerations together, it is hardly surprising that waste disposal has been rapidly privatizing in the United States while private toll road proposals have moved very slowly.

Attempts to develop new intercity rail passenger services in the United States through private efforts also illustrate some basic principles of privatization. This industry's experience is much like that of urban rail transit and for many of the same reasons. Intercity rail passenger service was initiated by private railroads in the nineteenth century, but with the onslaught of bus, automobile, and air competition in the twentieth century, profitability declined and passenger services were steadily abandoned. The decline was arrested in 1970 when a government-subsidized corporation, Amtrak, was created to take over and operate much of the remaining passenger services. Amtrak provides service over a relatively limited network and with a few exceptions (in the Washington–New York–Boston corridor) at top speeds of 80 miles an hour or less.

Beginning in the 1980s, private backers proposed building new high-speed (120 miles an hour or better) rail passenger lines in about a half dozen intercity corridors including Los Angeles–San Diego, Los Angeles–Las Vegas, Miami–Tampa, and Dallas–Houston.[12] Most of these propos-

12. For background on these proposals see Transportation Research Board, *In Pursuit of Speed: New Options for Intercity Passenger Transports*, Special Report 233 (Washington: Transportation Research Board, 1991).

als were eventually withdrawn. Basically, they failed the simplest of all private sector economic tests, that is, under almost any reasonable assumption they would not be profitable from user charges alone. Although some efficiency gains might be achieved by privatizing rail, these are not proven and probably not sufficient to carry most proposals over the threshold of being financially self-sufficient. The proposed Miami-Tampa line would have offset its rail losses with profits from real estate development along its right-of-way, for example. The state government was inclined to give permission for the proposed developments and exercise eminent domain on the project's behalf, but some local communities were less enthusiastic. In any event, backers withdrew their proposal when the Florida real estate market collapsed at the end of the 1980s. Intercity railroads are also likely to face many of the same serious environmental and distributional difficulties as new high-performance roads.

Airline privatizations illustrate the same principles. The international airline industry was once constituted almost exclusively (the United States being the major exception) of state-owned enterprises. Today that is no longer true as international airlines have been rapidly privatizing. International airline markets are, at minimum, characterized by duopoly and usually by substantial competition, in some cases even intense competition. Private airlines have almost always achieved lower costs than the state-owned international carriers, especially after adjustment for any national or geographic differences in labor costs. The major redistributions usually effected by privatization of international airlines are toward consumers, through a reduction of fares, and away from unionized labor forces (because of the introduction of more flexible work rules and, sometimes, lower wage scales); on the whole, these seem to be redistributions that are politically acceptable. Environmental problems are little affected one way or another by privatization, and international airline operations often can be profitable with the lowered cost structure that usually accompanies privatization.

Domestic airlines tend to be somewhat more difficult to privatize. First, domestic air operations are usually not as competitive as international (the United States again being a major exception); indeed, often domestic operations are well-protected monopolies, while in the worst case an international market will be a cartelized duopoly. Furthermore, redistribution or transfer problems are likely to be greater for domestic than international operations. Short-haul local air service, which characterizes much domestic air activity, tends to be subsidized, often being rendered at fares not only below the costs of the existing state-owned enterprise but also beneath the costs of any potential legitimate private alternative. These fare subsidies commonly are justified by governments as a means of achieving certain national integration goals, for example, tying isolated rural areas into the mainstream of national life, in short, still another example of the "ubiquitous standardized network" syndrome outlined earlier.

Intercity buses are prime candidates for privatization in both the developed and developing world, and indeed in many countries have never been part of the public sector. Competition can be effective, moreover, at almost all levels of economic development. In developing countries buses are often the principal form of intercity passenger transportation and economies of scale so modest that most corridors carry sufficient volumes to support several competing bus lines. In industrialized countries competition is more likely to come from other modes of intercity transport: auto, air and, often publicly subsidized, rail. The efficiency gains from privatization are likely to be great, as illustrated by the British experience of privatization in the early 1980s.[13] Profitability is generally not a problem, except possibly in highly developed societies like the United States. Distributional problems are usually rather more acute in less developed countries than in the developed; cheap intercity bus transportation is more likely to be viewed as a fundamental necessity or national goal since fewer alternative modes of transport are available and all except the wealthiest people in developing countries are likely to rely on bus transport for most of their intercity travel needs. The safety gains from public control of bus services are likely to be greater in the developing world than in the industrialized; in developed countries policing and other methods of insuring safe practices by private bus operators are usually more fully established.

Electricity generation, in contrast with electricity transmission or distribution, is likely to be a better candidate for privatization in the industrial countries than in the less developed. To start, industrialization creates more opportunities for alternative sources of electricity generation from cogeneration and similar activities, thereby heightening the potential competitiveness of such activities in industrialized countries and the possible efficiency gains. The low-cost universal availability of electricity (the ubiquitous standardized network goal) is, however, more likely to be still on the public policy agenda in less developed countries than in

13. See, for example, John Vickers and George Yarrow, *Privatization: An Economic Analysis* (MIT Press, 1988), pp. 366–75.

developed; the less developed country therefore may have more political incentive to retain complete control over the activity.

Water supply and sewage treatment, however, may be slightly more attractive candidates for privatization in less developed countries than in the developed. Competition is somewhat difficult to establish, since there is seldom a ready alternative source of supply or treatment for any given facility. Vendors can compete for a long-term construction and management contract for an entirely new facility or simply to manage an already existing publicly owned facility. In the developed countries some facilities will almost invariably be in place; these may not be state-of-the-art but nevertheless usually perform their tasks adequately. By contrast, many third world countries have only begun to cope with such problems. There is little existing investment and often therefore a need to create new "greenfield" facilities. In such circumstances an established private vendor, who has built such facilities several times previously, can bring experience and the newest technologies to the third world country at relatively low cost—certainly lower than the third world country itself is likely to achieve starting from scratch.

Some very impressionistic evaluations of the possibilities for achieving successful privatizations for various activities are shown in table 15-1. These are based on a composite evaluation of the basic characteristics—competitiveness, comparative efficiency, redistributions, externalities, and subsidy requirements—for each of the activities. Their tentative and impressionistic character cannot be overstated. They are meant to stimulate better alternative evaluations, and, above all, illustrate the applicability, or lack thereof, of the basic concepts.

The distinction between developed and developing countries used in table 15-1 is also obviously overly simple. As observed in part 2, when discussing highway privatizations, neither very undeveloped or highly developed countries are likely to be good candidates for private toll road developments; rather, countries in-between, particularly at the stage of early take-off into a modern consumer society, are likely to be the best prospects. Private commuter and transit rail might also enjoy their best prospects at such an intermediate stage of development.

Some Concluding Observations

In sum, privatization, especially when undertaken in a competitive sector and combined with deregulation, seems to have been beneficial

TABLE 15-1. Prospects for Privatization of Commonly Proposed Activities, Developed and Developing Countries

	Prospects for					
Activity and stage of development	*Competitive market*	*Large efficiency gains*	*Minimal transfers*	*Few externalities*	*Profitability from user charges*	*Overall success*
Solid waste disposal	Strong	Strong	Strong	Medium	Strong	Strong
Toll roads						
Developed countries	Medium	Medium	Low	Low	Low	Low
Developing countries	Medium	Medium	Low	Medium	Medium	Medium
International airlines	Strong	Strong	Medium	Strong	Strong	Strong
Domestic airlines (except U.S.)	Medium	Strong	Low	Strong	Medium	Medium
Urban rail transit (new lines)	Strong	Strong	Low	Low	Low	Low
Intercity passenger rail (new lines)						
Developed countries	Strong	Strong	Medium	Low	Low	Low
Developing countries	Medium	Strong	Medium	Low	Medium	Medium
Urban transit buses						
Developed countries	Medium	Strong	Medium	Strong	Low	Medium
Developing countries	Strong	Strong	Medium	Strong	Strong	Strong
Intercity buses						
Developed countries	Strong	Strong	Medium	Low	Medium	Medium
Developing countries	Strong	Strong	Medium	Medium	Strong	Strong
Airports	Medium	Medium	Low	Medium	Strong	Medium
Telecommunications						
Developed countries	Medium	Strong	Strong	Strong	Strong	Strong
Developing countries	Low	Strong	Medium	Strong	Medium	Medium
Electricity generation						
Developed countries	Strong	Medium	Strong	Strong	Strong	Strong
Developing countries	Medium	Medium	Strong	Strong	Strong	Medium
Electricity distribution						
Developed countries	Low	Strong	Medium	Low	Strong	Medium
Developing countries	Low	Strong	Low	Low	Medium	Low
Urban water supply						
Developed countries	Low	Medium	Medium	Medium	Strong	Low
Developing countries	Low	Strong	Medium	Medium	Strong	Medium

more often than not. The limited experience thus far suggests, however, that privatization may be helpful but is no panacea. Some shortfalls in public investment and service provisions may be well suited to a privatization solution while others may not be. What we do know with increasing certainty is that suitability will depend on several circumstances, above all the competitiveness of the markets served and the realizable extent of any efficiency gains; however, while less commonly mentioned, other factors—such as the extent and character of any redistributions and the complexity of environmental or externality issues—will also be important, sometimes even crucial. Other considerations will also be influential, such as the stage of economic development and the legal and political milieu.

This complexity and ambiguity should come as no surprise. Some basic issues of public policy are at play, especially that of defining the frontier between the public and private sectors. What should be their respective roles? What does the private sector do best? What does the public sector do best? Neither sector is likely to have an exclusive and total franchise in a civilized society. The privatization movement and its close kin of deregulation have taught us much about the merits and roles of the two sectors, but there is still much to learn.

Index